# Fusion Foodways of Africa's Gold Coast in the Atlantic Era

# Atlantic World

## Europe, Africa and the Americas, 1500–1830

*Edited by*

Benjamin Schmidt
*University of Washington*

and

Wim Klooster
*Clark University*

VOLUME 26

*The titles published in this series are listed at brill.nl/aw*

# Fusion Foodways of Africa's Gold Coast in the Atlantic Era

*By*

J. D. La Fleur

BRILL

LEIDEN · BOSTON
2012

*Cover illustration*: Maize Cobs Cast from Nature. Collection: Tom Phillips. Photograph: Heini Schneebeli.

Library of Congress Cataloging-in-Publication Data

La Fleur, J. D. (James Daniel), 1969–
Fusion foodways of Africa's Gold Coast in the Atlantic era / by J.D. La Fleur.
p. cm. — (Atlantic world : Europe, Africa and the Americas, 1500–1830, ISSN 1570-0542 ; v. 26)
Includes bibliographical references and index.
ISBN 978-90-04-22412-4 (hardback : alk. paper) — ISBN 978-90-04-23409-3 (e-book) 1. Food crops—Ghana—History. 2. Food habits—Ghana—History. 3. Agriculture—Ghana—History. 4. Plant introduction—Ghana—History. 5. Crops—Adaptation—Ghana—History. 6. Starch crops—Ghana—History. 7. Cassava as food—History. I. Title. II. Series: Atlantic world (Leiden, Netherlands) ; v. 26.

SB176.G4L3 20012
641.30309667—dc23

2012021561

This publication has been typeset in the multilingual "Brill" typeface. With over 5,100 characters covering Latin, IPA, Greek, and Cyrillic, this typeface is especially suitable for use in the humanities. For more information, please see www.brill.nl/brill-typeface.

ISSN 1570-0542
ISBN 978-90-04-22412-4 (hardback)
ISBN 978-90-04-23409-3 (e-book)



This book is printed on acid-free paper.

*For Dena, our daughters, and our family*

CONTENTS

# MAPS, ILLUSTRATIONS, AND WORD LISTS

## *Maps*



## *Illustrations*



## *Word Lists*

# PREFACE AND ACKNOWLEDGEMENTS

Every book has its own backstory, and bringing it before readers seems only fair. This project began in a frustrated search for information to fill a few footnotes in my master's thesis. At the time, I was translating and editing an Atlantic-era Dutch traveler's account of Africa (later expanded, revised, and published by the Hakluyt Society) and found myself confronted with all sorts of odd plant and food names which were certainly no longer used in modern Dutch, which could not be found in the Dutch equivalent of the *Oxford English Dictionary*, and which were not obvious loans from any other language in which I had some facility. So I thought to attack the problem from another angle. I would delve into the historiography of early Africa as seen through the excellent academic collection held by the University of Virginia's Alderman Library to try to establish a set of food crops known to have been grown in Africa in this era, and from this I would deduce by a process of elimination the identification of the mysterious food crops. But my review of the existing literature proved my alternative strategy to be equally infertile, as I discovered weaknesses and inconsistencies in the scholarly knowledge about—and lack of serious intellectual engagement with—the history of specific crops, farming strategies, agrarian knowledge, and the foods and culinary traditions of what was (and continues to be) overwhelmingly a population of farmers. And, of course, everyone eats, so the significance seemed only greater. I was also intrigued by the methodological challenge of researching and writing about the agricultural past of West Africa given the polyglot nature of the early documentation for this region and the absolutely critical role non-documentary sources would play in reconstructing and telling this history. I pitched the idea as a possible dissertation topic for the first time to a group of friends—our moveable feast of graduate students and faculty—during a half-time break from watching some 1994 World Cup game in the wonderfully air-conditioned basement of the house that our mates, Kenda Mutongi and Alan de Gooyer, were house-sitting for our professor-friend, Julia Clancy-Smith. Mine was an idea without a plan.

I count myself fortunate to have enjoyed so much help as I pursued a way. As an undergraduate and then, after a pause, a graduate student at University of Virginia, I learned so much from my teachers in ways I can still identify within this current project. They include, here in roughly the chronological order I first encountered them: Ben Ray, Carlos Eire, David Gordon White, Roy Wagner, Stephan Palmié, Armstead Robinson, Ed Ayers, Reginald Butler, Jeanne Maddox Toungara, Gertrude Fraser, Adria LaViolette, Robert Fatton, and John Mason. As a graduate student, I enjoyed a formative time of my intellectual life with outstanding peers: Emmanuel Akyeampong, Doug Chambers, Andy Lewis, Kelly Mulroney, Rebecca Popenoe, Anne Sarah Rubin, and Phil Troutman.

I moved to Leiden University (Universiteit Leiden) in 1996 to extend my research in the documentation for early Africa and to explore the potential of historical linguistics. The human talents, the intellectual atmosphere, and the research resources were so rich that I chose to stay a while. I could not have landed in better company than Felix Ameka, James Essegbey, Joscelyn Essegbey, Maarten Mous, Robert Ross, and Thilo Schadeberg. As I attacked my research problem from an historical point of view and noticed problems in the historiography of plantains, Gerda Rossel was bringing to fruition her study of the comparative botany and linguistics for this plant. Leiden was a wonderfully stimulating atmosphere, with talented and generous colleagues who helped me make continued professional progress even after I had finished my PhD, and began to weave teaching and research work around my commitment as a new father. Words fail to measure my appreciation.

The international community of scholars engaged in the research and writing of early African history is small but vibrant, rich with generosity, showing very little pomposity or territorialism, and fervently optimistic for Africa's future. I thank James Anquandah, Chris DeCorse, Wilhelmina Donkoh, Adam Jones, Robin Law, Tom McCaskie, Akosua Perbi, John Thornton, Larry Yarak, Ivor Wilks, and Selena Axelrod Winsnes, as well the memory of Albert van Dantzig and René Baesjou. Beyond the academy, I thank Emily Asiedu (of Kokomlemle, Ghana) for her hospitality as home-base during my fieldwork (and which my wife still charitably calls our 'honeymoon,' because it was) and honor the contribution of the late Ywan von Dewall (my neighbor in Leiden) to what I know about Dutch language and its history, and our great friendship.

Considerable institutional financial support came in the form of a Dupont Fellowship from the Graduate School of Arts and Sciences, University of Virginia (for study-abroad of Dutch as a foreign language); a Pre-Dissertation Fellowship from the Social Science Research Council and American Council of Learned Societies (for field research in Ghana); a Fulbright Fellowship administered by the Institute of International Education and the Netherlands-America Commission for International Exchange (for training and research in the Netherlands); a Dean's Dissertation Year Fellowship from the Graduate School of Arts and Sciences, University of Virginia; and some research funds from the College of William & Mary. I send my further thanks to the staffs of Alderman Library (University of Virginia), the National Library of the Netherlands (Koninklijke Bibliotheek, The Hague), the National Archives of the Netherlands (Nationaal Archief, The Hague), the Royal Netherlands Institute of Southeast Asian and Caribean Studies (Koninklijk Instituut voor Taal-, Land- en Volkenkunde, Leiden), the African Studies Centre (Afrika-Studiecentrum, Leiden), the National Herbarium of the Netherlands (Nationaal Herbarium Nederland, Leiden), and the Leiden University Library (Universiteitsbibliotheek Leiden) and its *Dousa-kamer* (special collections department).

During my time at William & Mary I have enjoyed good colleagues, drawn from the faculty and the students alike, who have encouraged my thinking about this project and African history. I thank in particular Jim Axtell, Ron Hoffman, Kris Lane, Paul Mapp, Ted Maris-Wolf, Neil Norman, Kim Phillips, Jeremy Pope, Allison Powers, Rich Price, Sally Price, Brett Rushforth, Carol Sheriff, Rob Vinson, and Chitralekha Zutshi.

Stan Alpern, my pen pal of so many years, helped with expert advice and generous attention to all manner of details on an early draft. I also benefited from the advice of several anonymous readers, as well as the recommendations of the Atlantic World series editors, Wim Klooster and Ben Schmidt. Tanya Buckingham and her team at the University of Wisconsin's Cartography Lab did a splendid job with the maps.

At my side throughout the entire project there has been the absolutely tireless Joe Miller, serving as my advisor as a graduate student (and beyond), my teacher, and then my colleague. But I am happiest to call him my friend.

The Dutch have an old workman's saying which I will translate with appropriate literary license as, 'the last bit of the load is always the heaviest to lift' (*de laatste loodjes wegen het zwaarst*). My thanks for helping shoulder the final load belong to Marti Huetink, Rosanna Woensdregt, and Rachel Crofut at Brill; and to my parents, Ken and Judy La Fleur, and my wife's parents, Ken and Carolyn Chapman, for complete support throughout the years. Their showering of love and devotion to their grandchildren created the time I absolutely needed to complete this project.

# NOTES ON LINGUISTIC EVIDENCE AND AFRICAN LANGUAGES

Although I have made considerable use of linguistic evidence in this study, I have not cited source materials for each bit of linguistic information in the body of the work. To do so, I feared, would encumber and distract the reader with dozens of repetitive notes at the foot of the pages. I have instead followed the scholarly convention in linguistic scholarship with regard to citation for such evidence, and therefore list my most frequently used source materials here below for individual languages and language groups. For historic (pre-1800 CE or so) language information cited in the text, sources are listed with full bibliographical detail in notes; and in the charts of vernacular names, they are marked by the collector's surname and the approximate date of collection.[1] Readers are entrusted to use the bibliography accordingly. No attempt has been made here to systematically change all source data from the multitude of orthographies in which they were originally recorded into a standardized one here. Modern linguistic information is presented here in the International Phonetic Alphabet and in font made available by the Summer Institute of Linguistics.

### *Gã-Adangme*


M. E. Kropp, *Comparative African Wordlists no. 2: Ga, Adangme and Ewe (Lomé) with English Gloss* (Legon: Institute of African Studies, University of Ghana, 1966).

M. E. Kropp Dakubu, *Ga-English Dictionary* (Legon: Institute of African Studies, University of Ghana, 1973).

M. E. Kropp Dakubu, "A Survey of Borrowed Words in Dangme," *Research Review*, supplement 4 (1973), 81–128.


### *Guang languages*


J. M. Stewart, *Comparative African Wordlists no. 1: Awutu, Larteh, Nkonya and Krachi with Glosses in English and Twi* (Legon: Institute of African Studies, University of Ghana, 1966).

Keith L. Snider, "Studies in Guang Phonology" (unpublished PhD diss., African Linguistics, Universiteit Leiden, 1990).


[1] Approximate dates of collection cited herein are consistent with those in J. D. Fage, *Original Sources for Precolonial Western Africa Published in European Languages*, 2nd ed. rev. (Madison: African Studies Program, University of Wisconsin, Madison, 1994).

### *Akan*

Rev. A. C. Crass, *Comparative African Wordlists no. 4: A Dictionary of the Chokosi Language* (Legon: Institute of African Studies, University of Ghana, 1970).

C. A. Akrofi and G. L. Botchey, *English, Twi-Ga Dictionary* (Accra: Waterville Publishing House, 1968 [rev. ed. 1980]).

C. W. Welman, *A Preliminary Study of the Nzima Language* (London: Crown Agents, 1926).

### *Ewe*

Diedrich Westermann, *A Study of the Ewe Language* (Oxford University Press, 1930).

Diedrich Westermann, *Wörterbuch der Ewe-Sprache*, 2 vols. (Berlin: D. Reimer, 1905–1906).

Diedrich Westermann, *Evefiala or Ewe-English dictionary*, 2 vols. (Berlin: D. Reimer, 1930).

C. A. Akrofi, G. L. Botchey, and B. K. Takyi, *An English, Akan-Ewe-Ga Dictionary* (Accra: Presbyterian Press, 1996).

### *Togo Remnant*

M. E. Kropp, *Comparative African Wordlists no. 3: Lefana, Akpafu, and Avatime with English Gloss* (Legon: Institute of African Studies, University of Ghana, 1967).

CHAPTER ONE

# FINDING HISTORY IN EARLY AFRO-ATLANTIC FOODWAYS

*Osikani na odi ade a ɛyɛ dɛ.*
'A rich man eats delicious foods.'

*Aduane panyin ne aburo.*
'The Chief among foods is maize.'

*Okuafo afɛre ɔbankye, na kɔtɔkɔnso afɛre.*
'The farmer is ashamed to eat cassava and the porcupine is also ashamed to eat it.'[1]

This project explores the cultural history of farming and food in the Atlantic era, focusing on the experience of people of the Gold Coast of West Africa (roughly equivalent to the modern nation of Ghana)[2] during the so-called 'Columbian exchange' of people, plants, animals, diseases, and ideas. The era roughly between 1500 CE and 1850 CE in Africa is commonly characterized as the 'era of the slave trade,' which reflects an older historiographical interest in Africa's external contacts, and in particular its export of captives to the Americas. But to describe this era in terms of what people in Africa did to produce those captives and what they did with what they received in exchange, that is the history of those (indeed the great majority) who remained on the continent, requires us to consider what the most significant changes were in *their* lives.

[1] J. J. Christaller (compiler), *Three Thousand Six Hundred Proverbs* (*from the Asante and Fante Language*) [originally published as *A Collection of Three Thousand and Six Hundred Tshi Proverbs* (Basel, 1879)], trans. Kofi Ron Lange (Lewiston NY: Edwin Mellon Press, 1990), 87, 150, 244.

[2] The independent African state of Ghana was established in 1957. Prior to independence, under British overrule, that area was know as the Gold Coast but was divided into three administrative regions: the Gold Coast, Asante, and Northern territories. After World War I, the British also administered a fourth (formerly German) area, Togoland, under an international mandate. This area was too incorporated into the independent state of Ghana. Throughout this study, the parameters of the 'Gold Coast' described remain consistent with the contemporary idea of the Atlantic era: that is, the area behind the coast lying between the Tano river (to the west, where Europeans first encountered substantial amounts of gold) and the Volta river (to the east, where European and African coast-hugging maritime efforts were interrupted by the river's strong flow). However, neighboring areas frequently merit mention in this study as they provide interesting contrasts with, or supporting elements to, the Gold Coast history described herein.

As most people in Atlantic-era West Africa—as in contemporary Europe and the Americas—were farmers, fields and gardens were the primary terrain where they engaged the opportunities and challenges of nascent globalization. Although historians in the past tended to limit African contributions to New World agriculture to their brute labor, there is growing documentation that slaves hailing from specific regions of Africa drew on the expertise and knowledge of particular crops and agrarian methods of their homelands to develop American plantations. But for so long as the history of agriculture in West Africa remained a neglected field of inquiry, historians inevitably viewed their African agents of American change as drawing on a 'traditional agriculture' rooted in immutable ethnic (or 'tribal') social structures—and described aspects of African-American foodways as inert 'survivals' without agents who adapted them to new circumstances in Africa and in the Americas.

Gold Coast agriculturalists reaped both physical and cultural sustenance from their gardens. The diets of Gold Coast farmers, like those of other agriculturists throughout world history, consisted of a limited range of carbohydrates, or 'starchy staple,' and a wide range of fruit, vegetable and protein supplements. Before *circa* 1500 African yams were the staple carbohydrates in the forest zone of West Africa, and sorghum and millets (all of them native African grains) were staples in savanna patches north of the forest and the grasslands stretching along the southeastern parts of the coast. These agricultural products were supplemented by other forms of vegeculture, meaning the careful tending and harvesting of plants such as the palm or various nut trees that propagated themselves without human cultivation. Fishing, foraging, hunting, and gathering supplemented farmers' diets at crucial periods in the agricultural cycle when planted food crops were yet immature or when gardens needed little attention. Farming, though, was the focus of their strategy. The annual cultural calendar of initiations, celebrations, ascensions, and burials shadowed the agricultural calendar of cutting and clearing land, planting crops, weeding gardens, and harvesting and storing produce of the starchy staples. To organize the specific forms of labor needed at different periods of the staple crops' cycles, farmers invented social relationships around ideologies of gender, kinship, clientage, and slavery.[3] The replacement of these staples by new starchy food crops therefore marked a revolution.

[3] Charles Piot, "Production, Exchange and Symbolic Forms among the Kabre," (Unpublished PhD diss., Anthropology, University of Virginia, 1986); J. P. Chrétien, "Les années de

For heuristic purposes this study focuses on the new carbohydrates that the farmers of the Gold Coast region adopted: maize and cassava (manioc) from the Americas, and plantains and new varieties of yams and rice originally from Asia. By extension, this project also concerns significant non-agricultural strategies to procure food, such as hunting, gathering, foraging, fishing, livestock-raising, and trade, as well as the cultivation of other new fruits and vegetables they adopted in those same years.

### *Grist for the Mill: Africanist Historiography of Pre-Colonial Agriculture*

The history of African farming and food is the mysterious 'black box' of all of early African (and, by extension, Atlantic) historiography. African historians long made problematic assumptions about people and their crops and foods in this era, and these assumptions raised but did not resolve important historical questions. Assumptions about the agricultural past reach to the basic premises of early African historiography by forming the bases for arguments about demographic history, labor (slavery) history, the history of technology, climate history, environmental history, the history of social systems used to procure labor, and the history of state systems devised to extract labor and its primarily agricultural product. However, since these studies treat farming and food as a sub-field of political economy they neglect the intimate relationship between farmers and their crops, between agriculture and culture, between raw produce and cuisine, and between farmers' work and 'harvest home' festivities they celebrated when the staple crops were finally in. Thus Africa's agricultural history has been engaged more as 'grist for the mill' in historiographical argumentation rather than as independent subject worthy of serious, sustained inquiry into agrarian knowledge, changing crops regimes and cultivation methods, and cultural ideas about specific food crops and cuisine.

Typically historians have assumed that the botanical qualities of crops determined changes in the lives of those who grew and ate them. Thus historians infer transformations in agricultural societies from the qualities of their changing crops. Historians have focused their attention also

l'éleusine, du sorgho et du haricot dans l'ancien Burundi ecologie et idéologie," *African Economic History* 7 (1979), 75–92.

on the American starches—maize and cassava—that they thought were the only new crops of the Atlantic-era. The physical requirements and nutritional values of these food crops are thoroughly documented, though through the crops' modern variations. For example, maize can withstand high, tropical humidity better than the old African cereals. This tolerance allows maize to thrive on the 'wet' savannas and in forest clearings. Maize is also much quicker to fruit than either sorghum or millets and therefore can be planted twice a year, to take advantage of the tropic's two rainy seasons, thus producing significantly higher yields than the older grains. But maize is consequently also very sensitive to drought, and without rainfall it will not fruit to maturity. Maize also exhausts light tropical soils quickly, leaving garden plots depleted of nitrogen and other soil-nutrients.[4] Cassava, on the other hand, is a tuber and grows well on even the poorest, nutrient-deficient soils and is well adapted to the forest. Cassava yields are much greater than either yams or bananas, and its leaves are a nutritious vegetable too. Its roots do not need to be harvested at a fixed date (in contrast to grains and yams) and can be left underground for several years without rotting. Cassava roots are high in calories but are of little nutritional value. Several species, moreover, contain poisonous cyanogenic glycosides, although these can be removed by laborious heating and fermenting processes.[5]

The history of human interaction with these highly specific soil, labor, and processing requirements and crops' nutritional values could, prior to the current study, only be speculated for West Africa's Gold Coast. Historians of pre-colonial Africa have generally been interested only in a single aspect of these new crops: their (presumed) higher annual yields. This quality of the new crops was widely acclaimed to be the single benefit bestowed upon Africa during centuries of ruinous slaving.[6] The demographers argued that the American crops improved Africans' diets with

---

[4] On the plant biology of the crops discussed above, see Major M. Goodman, "Maize (*Zea mays*)," in N. W. Simmonds, ed., *Evolution of Crop Plants* (New York: Longman, 1976), 128–36; J. W. Purseglove, "The Origins and Migration of Crops in Tropical Africa," in Jack Harlan and Jan M. J. de Wet, *et al.*, *Origins of African Plant Domestication* (The Hague: Mouton, 1976), 291–309; Marvin P. Miracle, *Maize in Tropical Africa* (Madison: University of Wisconsin Press, 1966).

[5] William O. Jones, *Manioc in Africa* (Stanford: Stanford University Press, 1959); D. L. Jennings, "Cassava (*Manihot esculenta*)," in Simmonds, ed., *Evolution of Crop Plants*, 81–4.

[6] *Contra* Walter Rodney, *How Europe Underdeveloped Africa* (London: Bogle-l'Ouverture Publications, 1972), 111–2.

increased calories, which curbed mortality and enhanced female fertility.[7] Moreover, some demographers argued that subsequent regional population growth may have been sufficient to offset losses to the Atlantic trade. Philip Curtin proposed—speculatively, as was, and still remains, appropriate to a field untilled by serious research—this in his 1969 *Census*: 'it seems possible and even probable that population growth resulting from new food crops exceeded population loss through the slave trade.'[8] More recently, another historical demographer, Patrick Manning, repeated this unexamined stress on New World cultigen productivity and historical demography but used it to support precisely the opposite conclusion. He pointed to lopsided sex and age ratios to show that factors such as the new food crops 'were not powerful enough nor variable enough to overwhelm the seriously negative and distorting impact of slave exports on the African population.'[9] The shortcoming of such perspectives is that they assume social homogeneity to construct demographic aggregates. Manning's study, for example, differentiates only in bio-demographic terms such as sex and age. The demographers suggest that the cultivation and consumption of New World starchy crops was undertaken immediately and was widespread and uniform, both socially and geographically.

West Africanists also explained pre-colonial state-formation as a similarly unexamined result of agricultural development and demographic growth. Ivor Wilks explained the emergence of the kingdom of Asante in the late seventeenth century as the culmination of a three century-long transformation from hunting and gathering to agricultural production.[10] What made Asante notable was that it lay in a tropical forest zone, which was then widely understood by theorists to be hostile to large-scale

---

[7] Population growth driven by agricultural innovation, crop diversification and resultant surplus is a standard model borrowed from European historiography; see for example, William L. Langer, "Europe's Initial Population Explosion," *American Historical Review* 69, 1 (1963), 1–17; E. A. Wrigley, *Population and History* (New York, 1969). Esther Boserup argued the exact opposite model of causation in tropical areas, suggesting that farmers increased food production to meet the demands of growing populations; in Esther Boserup, *The Conditions of Agricultural Growth* (Chicago: Aldine, 1965).

[8] Philip D. Curtin, *The Atlantic Slave Trade: A Census* (Madison: University of Wisconsin Press, 1969), 270.

[9] Patrick Manning, *Slavery and African Life* (Cambridge: Cambridge University Press, 1990), 59.

[10] Ivor Wilks, "Land, Labor, Capital, and the Forest Kingdom of Asante: A Model of Early Change," in J. Friedman and M. J. Rowlands, eds., *The Evolution of Social Systems* (Pittsburgh: University of Pittsburgh Press, 1978), 487–534 [revised and reprinted under the same chapter title in Ivor Wilks, *Forests of Gold* (Athens: University of Georgia Press, 1993), 40–90].

political amalgamation. Wilks explained that the Akan (explained as a generalized linguistic-cultural group) had achieved the density of population needed to spur the development of a complex political hierarchy by selling the gold they mined for slaves. They then sent these slaves into the fields to produce food and during seasons of low demand for agricultural labor to the pits to mine gold. For Wilks, the major utility of the slaves to the ascent of Asante was that they provided the heavy labor needed to clear the forest for agriculture, as needed to produce food for future demographic expansion.[11] But Wilks did not look to see that Asante farmers did not immediately and uniformly abandon their old, dependable foodways for new, yet-untested ones, nor that the foreign slaves brought with them their own ideas and talents with producing food and they thus significantly contributed to the historical development of the very institutions he targeted for analysis. Wilks's idea proved very durable (to his surprise, because he privately thought his argument about agriculture was more of a working model, and one which he continued to update as recently as 2005), and so subsequent historians embellished it with new insights from agricultural sciences and environmental history.[12] But none challenged its basic assumptions about the chronology of crop repertoires or the sometimes anachronistic evidence behind the model's argument.[13]

Ray Kea's *Settlements, Trade, and Polities on the Seventeenth-Century Gold Coast* credited the introduction of highly productive New World food crops (with emphasis on maize) with supporting state-formation, south of Asante, in the central region on the Gold Coast. His neo-marxist interpretation invoked early surplus grain production and subsequent demographic growth, urbanization, professional specialization, class formation, élite political domination of peasants and the appropriation of the increased rural reproductive potential.[14] Although agricultural production was central to Kea's reconstruction in an abstract, undifferentiated way, the introduction of specific new crops was not. He discussed the product

---

[11] Wilks, "Land, Labor, Capital," 488, 503–9.

[12] Conversation with Wilks (6 December 2002). He revised the model in 1993 (see note above) and again in 2005; see Ivor Wilks, "The Forest and the Twis," *Journal des Africanistes* 75, 1 (2005), 19–75.

[13] James C. McCann, *Green Land, Brown Land, Black Land: An Environmental History of Africa, 1800–1990* (Portsmouth NH: Heinemann, 1999), 109–140; James C. McCann, *Maize and Grace: Africa's Encounter with a New World Crop* (Cambridge MA: Harvard University Press, 2005), 23–58.

[14] Ray Kea, *Settlements, Trade, and Polities on the Seventeenth-Century Gold Coast* (Baltimore: Johns Hopkins University Press, 1982), 12, 16, 43–6, 48; see also his Map 1.2 which shaded the entire coast as 'areas of agricultural surplus.'

of farm labor, or food as commodity, which he described consistently only as an abstracted *millie*, 'a generic term used to designate maize, millet, and Guinea corn [=sorghum].'[15] Walter Hawthorne's description of the Atlantic-era development of agriculture by the Balanta in the marshy ground of Upper Guinea (now, Guinea-Bissau) acknowledges diversity in the grains that farmers sowed, but avers exploring in an historical way the substantive differences these farmers must have recognized between their familiar African rice and the newly introduced Asian species.[16] Here too, the productivity of the alien crop is invoked to explain demographic and demographically linked institutional development cooperative with the export of captives.

Demographic perspectives and demographically based political arguments, however, tell us little about 'pre-colonial' farmers, although indeed most people were farmers who spent most of their time working the land to produce the food that they ate. Better work has been done for pre-colonial Central and Southern Africa, where historians related the significance of agricultural history to more popular 'politics and trade' themes for early Africa. Joseph Miller connected pulses of South Atlantic slave exports to agricultural failures that turned savanna farmers on sandy Angolan soils into famine refugees, Imbangala warriors, and Brazilian slaves.[17] Elizabeth Eldredge has similarly explored nineteenth-century Zulu state formation and the violence of the so-called *mfecane* (Zulu, 'crushing') as the product of increasing inequalities between and within South African societies amid environmental and, thus, agricultural crisis.[18] Eldredge's work on early Lesotho develops in further detail the conflicts within societies that drought inevitably exacerbated. Central to her argument was Indian anthropologist Amartya Sen's theory of entitlement and the social causes of hunger, which she used to demonstrate that the BaSotho elite restricted 'legal' access to food amid

15 So described, parenthetically, in the endnotes to Kea, *Settlements, Trade, and Polities*, 339 n. 31.

16 Walter Hawthorne, "Nourishing a Stateless Society during the Slave Trade: The Rise of Balanta Paddy-Rice Production in Guinea-Bissau," *Journal of African History* 42, 1 (2001), 1–24; Walter Hawthorne, *Planting Rice and Harvesting Slaves: Transformations along the Guinea-Bissau Coast, 1400–1900* (Portsmouth NH: Heinemann, 2003).

17 Joseph C. Miller, "The Significance of Drought, Disease, and Famine in the Agriculturally Marginal Zones of West-Central Africa," *Journal of African History* 23, 1 (1982), 17–61.

18 Elizabeth Eldredge, "Sources of Conflict in Southern Africa, c. 1800–30: The 'Mfecane' Reconsidered," *Journal of African History* 33, 1 (1992), 1–35.

mild scarcity to create increasingly inegalitarian social relationships.[19] However, these studies treated agriculture as a sub-field of political economy and thus neglected the creative and intimate relationship between farmers and their crops, between agriculture and culture, between farmers' work and the celebratory festivities they enjoyed when the staple crops were finally in. Elias Mandala, the historian of colonial south-central Africa (now, Malawi) lamented the tendency in African historiography to write about farmers—but not farming itself.[20]

## *Directions*

This book partakes in an emerging body of research that, for the first time, places farming and food at the center of histories of African communities in the Atlantic era. Though this new research is being pursued by scholars working largely independently of one another, and though these scholars remain very much intellectually engaged in local (or ethnic, or national) historiographies, our common challenge of working from scarce and problematic sources gives rise to shared questions about the historical relationship between people and plants. We share much in common, including overlapping tactics and mixed success in deploying: first, non-documentary methodologies capable of 'finding food' but which obscure human agency; and second, scientific (botanical, nutritional, and environmental) information that is sophisticated for the plants that people exploited but takes no account of the cognitive and cultural contexts in which farmers grew them, cooks prepared them, and consumers ate them; and third, often richly textured but fundamentally anachronistic modern ethnographic insights.

This new, emergent scholarship draws on a long established and enduring intellectual framework for the Atlantic world. The interdisciplinary model of globalization of biological resources in this era is familiar to all of us from Alfred Crosby's *Columbian Exchange* and his *Ecological Imperialism*, and Jared Diamond's *Guns, Germs, and Steel*.[21] African historians

[19] Elizabeth Eldredge, "Drought, Famine, and Disease in Nineteenth-Century Lesotho," *Journal of African History* 16, 1 (1987), 61–93; Amartya Sen, *Poverty and Famines: An Essay on Entitlement and Deprivation* (Oxford: Oxford University Press, 1981).

[20] Elias C. Mandala, *Work and Control in a Peasant Economy: A History of the Lower Tchiri Valley in Malawi, 1859–1960* (Madison: University of Wisconsin Press, 1990), 8–10.

[21] Alfred W. Crosby, The Columbian Exchange: Biological and Cultural Consequences of 1492 (Westport, Conn.: Greenwood, 1973); Alfred W. Crosby, *Ecological Imperialism: the*

inspired by this model fully tallied the inventory of biological resources new to the continent in the Atlantic era: farmers of West and Central Africa supplemented their cultivation and consumption of native 'starchy staples'—yams, millet, sorghum, and an indigenous species of rice—with new carbohydrate-rich foods including: maize and cassava from the Americas; and new varieties of yams and rice originally from Asia; and in West Africa, plantains (also known as cooking bananas), too, were a new crop. Africans also adapted and adopted many new fruits and vegetables: peanuts (also known as groundnuts), common (or kidney, haricot) beans, chile peppers, pumpkins, squashes, citrus fruits, pineapples, papayas, guavas, tiger nuts, tomatoes, onions, sugarcane, and the addictive non-food Atlantic contribution, tobacco. And there is yet more.[22] This was an expansive story—or more properly, multiple stories—varied and distinctive on the most local scale.

African historians who, at the time of publication of Crosby's introduction of the idea of the 'Columbian Exchange' in 1972, were working primarily on the political economy of pre-colonial states and on those states' capture of people for export to the New World borrowed directly from the Crosby's model to propose an indigenous, 'second' agricultural revolution in Africa prior to European overrule in the late nineteenth century. Doing so, these historians were able to reject the racist colonial ideology of the twentieth century that Africans were suspicious of technology, adverse to change, and needed to be forcibly pried out of unproductive, so-called 'traditional' agriculture by enlightened outsiders into progressive agronomy with saleable surpluses of foodstuffs, the cultivation of inedible cash-crops and, everywhere, the application of petrochemical fertilizers.[23]

Historians who found dynamism in Atlantic-era African agriculture thereby moved the horizon of 'traditional' foodways back into the era of unquestionably absolute African autonomy—prior to colonization, and preceding the Atlantic slave trade, to the era just before the fifteenth-century arrival of a few Portuguese caravels bearing botanical specimens.

*Biological Expansion of Europe, 900–1900* (Cambridge UK / New York: Cambridge University Press, 1986); Jared Diamond, *Guns, Germs, Steel: The Fates of Human Societies* (New York: Norton, 2005).

[22] For a regional survey, see Stanley Alpern, "The European Introduction of Crops to West Africa," *History in Africa* 19 (1992), 13–43; updated in "The Exotic Plants of Western Africa: Where They Came From and When," *History in Africa* 35 (2008), 63–102.

[23] J. D. La Fleur, "Plants: Imported Species," in John Middleton and Joseph Miller, eds., *New Encyclopedia of Africa*, 2nd ed. (New York: Scribner, 2008), vol. 4, 158–59.

But they confronted methodological problems that prevented them from fully recognizing Africans' agency in Atlantic-era transformations in farming, preparing, and eating food. They found nothing in the documents penned by literate foreign visitors to Africa that detailed the introduction of the new crops: not the personal identity of the historical agents, their precise physical locations and dates, their various motivations, or their methods. Historians searching contemporary documents for the very earliest mentions of the foreign crops in Africa confronted plant names that were so different from modern ones and really so vague that they offered scholars very wide latitude, but little certainty, in identifying them. In one famous—or notorious—example, this lead to the claim that maize was widely grown in Africa before 1492. Historians who listened to oral traditions in hopes of finding perspectives unpolluted by outside influences were disappointed to hear precious little about farming in the Atlantic era despite the historians' certainty that this agricultural revolution must have been so historically significant that it warranted remembering.

With archaeology and historical linguistic study still undeveloped for tropical Africa, historians turned—or returned, ironically enough—to the colonial-era theme of scientific progress. They explained that Africans adopted the new crops because these farmers were *rational*. Historians drew on colonial data of annual yields to show that the American starches were more productive (but in which terms, be it land, labor, other costs, or nutrition, they rarely questioned). The productive quality of the new crops was widely acclaimed to have been the single benefit bestowed upon Africa during centuries of ruinous slaving. Historians used the unquestioned productivity of the alien crops to suggest a series of successive conclusions: enhanced food supplies fueled demographic growth, urbanization, professional specialization, class formation, élite political domination of peasants, the appropriation of the increased rural reproductive potential, and pre-colonial state-formation. Barring the sort of detailed data for Africa that William Langer used to document similar transformations in Europe after the adoption of New World crops (particularly the potato),[24] the Africanists' hypotheses remained untested by contemporary evidence. But their approach to the relationship between people and plants was clear: they assumed that the botanical qualities of

[24] William L. Langer, "American Foods and Europe's Population Growth 1750–1850," *Journal of Social History* 8, 2 (Winter, 1975), 51–66; and later reworked as William H. McNeill, "How the Potato Changed the World's History," *Social Research* 1, 66 (1999), 67–83.

crops determined changes in the lives of those who grew and ate them. Historians inferred transformations in agricultural societies from the scientific qualities of their changing crops.

Africanists who had worked this field intensively in the 1970s then let it go fallow in the 1980s, as they and the new generation of young colleagues they trained increasingly turned their attention to the more recent African past, to colonial and post-colonial stories. What then in the past recent years has stimulated new energy in researching and writing about African foodways in the Atlantic era? For sure, some of us read public pleas and heard encouragement from senior Africanists, like Jan Vansina[25] and James McCann,[26] to take the *history* of agriculture seriously. And some must have been impressed by senior Americanists such as Ira Berlin and Philip Morgan, who together lamented that agriculture was so central to American slaves' experiences that it had often been overlooked and taken for granted.[27]

Our awareness of Atlantic scholarship has had a profound effect in directing new research. We were aware that although historians in the past tended to limit Africans' contributions to New World agriculture to their brawn (rather than their brains), that there was growing documentation that slaves hailing from specific regions of Africa drew on the expertise and knowledge of particular crops and agrarian methods of their homelands to develop American plantations. Particularly emphasized, celebrated, and controversial was work connecting West African rice farmers to the development of Low Country economies and societies.[28]

African historians who sought to complement Americanists' interests in early modern African agriculture took advantage of increasingly abundant scholarship from sibling disciplines—archaeology and comparative-historical linguistics, but also environmental science and geography—that, like history, are concerned with the past. These methods and sources

[25] Jan Vansina, "Finding Food and the History of Precolonial Equatorial Africa: a Plea," *African Economic History* 7 (1979), 9–20.

[26] James C. McCann, "Agriculture and African History," *Journal of African History* 32, 2 (1991), 507–13.

[27] Ira Berlin and Philip D. Morgan, eds., *Civilization and Culture: Labor and the Shaping of Slave Life in the Americas* (Charlottesville: University Press of Virginia, 1993).

[28] Judith Carney, *Black Rice: The African Origins of Rice Cultivation in the Americas* (Cambridge MA: Harvard University Press, 2001); David Eltis, and Philip Morgan and David Richardson, "Agency and Diaspora in Atlantic History: Reassessing the African Contribution to Rice Cultivation in the Americas," *The American Historical Review*, 112, (December 2007), 1329–1358; and colloquy in "The Question of Black Rice," *American Historical Review* 115, 1 (2010), 123–171.

moved historical inquiry of Afro-Atlantic foodways far beyond the chronological reach and geographical awareness of outsiders' documentation and thus allowed historians to finally expose 'traditional agriculture' for the ahistorical fallback it had always been, for want of the evidence accessible through these other scholarly disciplines. What we saw in the new evidence instead was *change* extending into the deep past. For example, Edda Fields and Walter Hawthorne showed that the 'classic' rice farmers of Upper Guinea who brought their skills to Americas made this crop their staple only relatively recently, probably in the Atlantic era.[29] They show that farmers seeking to make this change drew on (domestic) regional networks linking the coast with the Inland Niger Delta in the deep interior to acquire seeds and cultural knowledge.

## *Investigative Themes*

This project uses three themes to explore the history of West Africans' foodways. The first is the processes of farmers' technical adaptation of both old and new Atlantic crops. How did farmers adapt the crops to their preferences through selective breeding and the invention of new horticultural techniques, cooking technologies, and culinary forms? And how did they conceptualize these historical processes? The second is the cultural and social implications of implementing new Atlantic food technologies. How did farmers adapt to the highly specific soil, moisture, and processing requirements of the new staples, as well as their different nutritional values? How can the cultivation and consumption of particular crops help explain such social identities as ethnicity, gender, seniority, and wealth, given that people often expressed these in terms of prohibitions and privileges in the production and consumption of food (recalling the popular notion in so many languages that 'you are what you eat')? The third is the environmental history of the area. How did people adapt their productive strategies amid larger, long-term processes of climate changes and how did their changing efforts create cumulative changes in the landscapes where they made their living? How did their perceptions of their environments inform their actions?

---

[29] Edda Fields, *Deep Roots: Rice Farmers in West Africa and the African Diaspora* (Bloomington: Indiana University Press, 2008); Walter Hawthorne, *From Africa to Brazil: Culture, Identity, and an Atlantic Slave Trade, 1600–1830* (Cambridge UK: Cambridge University Press, 2010).

The remainder of present study is divided into four, chronologically progressive chapters. This essentially narrative strategy was necessary in order to create a basic chronology of people and their food plants. The next chapter begins the historical narrative in the very remote past with the earliest inhabitation of West Africa by foragers, and proceeds through their subsequent efforts to feed themselves and others there prior to the Atlantic era. Their efforts at subsistence generated a dynamic process directed towards solving a dialectical series of historically specific problems, and were not an inevitable development toward progress. The early inhabitants' motivations to adapt their food strategies to include yams, millets, and sorghums produced some unforeseen consequences—not all of them beneficial to all—including, most significantly, additional labor requirements in procuring raw starchy ingredients and preparing these into food. The resulting cumulative specialization of communities toward distinctive foodways and the concomitant trade among the diversified groups they created to maintain them undergirded the growth of ethnic distinctions that paralleled the environmental niches they had chosen to exploit and that structured complimentary exchange of foodstuffs—and other produce, including (but less often) people—across them.

This chapter also serves to introduce the reader to the foragers, farmers, herders, and fisherfolk of the area and the environments they inhabited. Both people and places are treated in a strictly historical manner. That is the early inhabitants are described in the context of *change* rather than in static terms as 'traditional' societies, and are presented without the typologies (Early Stone Age, etc.) archaeologists use to communicate epochs of change to one another but historians overhear as essentially static typologies;[30] and the physical environment they inhabited is also described as under change, both from global ecological forces and from the early foragers' and farmers' efforts to domesticate the landscape.

The impetus to write about the 'deep' past was borne, firstly, of intense personal curiosity about the greater historical context in which Gold Coast farmers involved themselves in the pan-Atlantic experimentation with new crops, agricultural strategies, and culinary styles. It seemed plain that Gold Coast farmers brought their own complex notions about food and feeding with them in their engagement of the Atlantic world and that

[30] Jan Vansina, "Historians, Are Archaeologists Your Siblings?," *History in Africa* 22 (1995), 369–408; Peter Robertshaw, "Sibling Rivalry? The Intersection of Archaeology and History," *History in Africa* 27 (2000), 261–86.

such ideas were historically dynamic. The sort of synchronic 'historical baseline' seen universally in the monographs penned by the first generation of Western academics to study Africa and against which I could contrast later, Atlantic-era developments was but an intellectual construction of limited value for my study.

The third chapter begins in the late-fifteenth century, when farmers around the three early Portuguese strongholds on the coast added new crops to their repertoires of agricultural experimentation. They attempted to cultivate some new food crops from the palette of crops that the Portuguese collected from throughout its worldwide empire in order to provision themselves, but local farmers stole the new crops to sow their own seeds of change. African farmers—arguably women—were experienced, hence cautious and selective. They knew how much was at stake and therefore tested the novelties in their kitchen gardens and adapted them to their needs and tastes before eventually adopting as staples the ones they found attractive. This perspective runs counter to the current understanding of an immediate and mechanistic 'revolution' based on the high-yielding (and therefore superior, it is assumed) American starches, maize and cassava. This chapter focuses on local farmers' deliberation in adapting and collectively adapting to only certain foreign starches—particularly maize and plantains, but also Asian rice—and the considerations they took into account in creating new cultures that supported them, thus to analyze their historical development of new food strategies, ethnic identities, cultural calendars, and gender distinctions. This chapter argues against the long-dominant historical assumption that plantains were grown and consumed on the Gold Coast prior to Portuguese maritime introduction. Linguistic and documentary evidence links the introduction of plantain culture to the importation of African slaves from several thousand miles away in Central Africa, where farmers had made these starchy fruits their staple crop. Less significant foreign agricultural resources adapted and then adopted in this era were sweet potatoes, chile peppers, and tobacco (from the Americas); sugarcane and citrus fruits (from the Mediterranean); coconuts (South Asia); and assorted European greens and European breeds of chickens and swine. This chapter places emphasis on the social, cultural, environmental contexts in which Gold Coast farmers experimented with their new starchy food crops. It places emphasis on the human experience of experimentation rather than the search for the earliest mention of specific exotic crops in Western Africa as noted in European sources under ambiguous and disputable contemporary names.

The fourth chapter follows the frontier of agricultural experimentation as farmers throughout the area struggled to survive the era of intensive slaving and slave exports. Local farmers repeated the process of trying out new plants in their kitchen gardens, selecting appropriate ones, and adapting them to their own preferences and the natural environmental limits of their garden plots and the constraints of climate irregularity and often violent competition to control the fruits—political as well as botanical—of contact with the Atlantic. This chapter examines the diversity of local agricultural and culinary adaptations through time, as farming communities made the new starchy crops their primary staples, and the unforeseen and unintended socio-economic, dietary, and cultural consequences of shaping these new foodways.

The fifth chapter emphasizes the rise of cassava farming from the era of the first slave exports to the mid-nineteenth century. Chronologically, this chapter overlaps somewhat with the previous one, but the weight of the chapter rests on the nineteenth century, when subsistence farmers everywhere eventually resorted to cassava cultivation in their struggle to create food security amid conditions of violent disorder and environmental degradation. Farmers were attracted to cassava's drought tolerance, general disease and pest resistance, and its ability to endure underground storage for long amounts of time in drier soils and surviving in less arable, but more defensible, terrain. Moreover they could grow the crop on soils too exhausted or too poor for any other crop, thereby alleviating the need for (male) labor to prepare new gardens. Although many varieties of cassava were introduced from the Americas, local farmers (probably women, as it was a crop disparaged by men) selected against the higher-yielding types that required more work to remove its potent toxins. Women preferred the 'sweeter,' less poisonous varieties that they could prepare into edible food more easily. Nowhere on the Gold Coast did male-dominated communities celebrate cassava harvesting, nor did they bestow male esteem on its consumers. In local proverbs, people poked fun of cassava as food that even the lowly porcupine was 'ashamed to eat,' but they also ridiculed farmers who did not plant some of the crop as insurance against the persistent threat of famine. The poorest, most war-ravaged farmers of the drought-stricken area east of the Volta river captured this crop's critical value to their own survival when they designated it (in the Ewe language), 'there is life,' with the implication, 'but little else.'

This study concludes with a sixth chapter setting the agricultural history of the Gold Coast in the Atlantic era in the larger intellectual context of food studies, a field largely grounded in anthropology, and in the field of 'Atlantic' history, which has had some trouble integrating Africa. Stressed here is the unique potential of applying historical emphasis on change to crops and cuisine as products of human adaptation, their significance as human culture as well as biological cultigens, and the high yields of conceptualizing the history of West African foodways in the contexts of the multiple and multi-directional biological, economic, and cultural cross-currents of the Atlantic era.

This study prompts a serious reappraisal of Atlantic models (notably Crosby's and Diamond's crop- and animal-structured notions, seen also in earlier Africanist historiography) in rigorously *historical* terms—that is, as humanistic change. Or, in another phrasing, scholars should revise their understandings of Africans' engagements with the Atlantic world from mechanistic 'adoption' to reveal the considered and contested adaptations by Africans living amid longer-term rhythms of change and long-standing involvement with inter-regional systems of human communities. The most important 'seeds of change' of the Atlantic era are not to be found in the DNA of crops and critters carried across the seas, but instead in the tradition of creativity and innovation of the people who engaged the challenges and opportunities of early globalization.

CHAPTER TWO

# INTRODUCING THE LAND TO CULTURE: AN INTERPRETATION[1]

Many tens of thousands of years ago, the first anatomically modern human hunters and gatherers settled the lands lying about the Volta river [see Map 1]. Establishing even just rough dates for the earliest archaeological evidence of human occupation in this area is terribly problematic. A few crude stone tools have been found and attributed with Oldowan and Acheulian antiquity (i.e., dating before 60,000 years ago), but the dating and typing remains very controversial.[2] The oldest human skeletal remains in tropical West Africa date from 11,000 years ago, at Iwo Eleru (Nigeria), and from less than 4,000 years ago for the area west of the lower Volta, but these dates seem much too recent for the first exploration of this area.[3] Whenever they arrived, the first foragers were not the last. The occupation of the area was not a single, dramatic 'event' but rather a slow, piecemeal process including many histories of movement, adaptation, and further movement and adaptation in the deep past. When these various groups of people arrived, they did not come unprepared. Rather, they all drew on an ancient hunting and gathering tradition, derived from the earliest human times on the East African savanna, to make their living here.

---

[1] The subtitle of this chapter is intended to make explicit the necessarily speculative character of the following interpretation. Intellectual inspiration for this analysis draws on the argument for just such a 'conjectural history,' in T. C. McCaskie, "Accumulation, Wealth, and Belief in Asante History," *Africa* 53, 1 (1983), 23–43.

[2] For the claims of great antiquity, see: Oliver Davies, *The Quaternary in the Coastlands of Guinea* (Glasgow: Jackson, 1964); Oliver Davies, *West Africa before the Europeans* (London: Methuen, 1967); Adu Boahen, "Ghana before the Coming of Europeans," *Ghana Social Science Journal* 4, 2 (1977), 96; Bassey Andah, "Early Paleolithic in West Africa: The Case of the Asokrochona Coastal Region of Accra Ghana," *West African Journal of Archaeology* 9 (1979), 47–84. For criticism, see: S. E. Nygaard and M. R. Talbolt, "Stone Age Archaeology and Environment on the Southern Accra Plains, Ghana," *Norwegian Archaeological Review* 17 (1984), 19–38; James Anquandah, *Rediscovering Ghana's Past* (London: Longman, 1982), 49–50; Ann Brower Stahl, "Innovations, Diffusion, and Culture Contact: The Holocene Archaeology of Ghana," *Journal of World Prehistory* 8, 1 (1994), 59.

[3] P. Allsworth-Jones, "The Earliest Human Settlement in West Africa and the Sahara," *West African Journal of Archaeology* 17 (1987), 87–128.

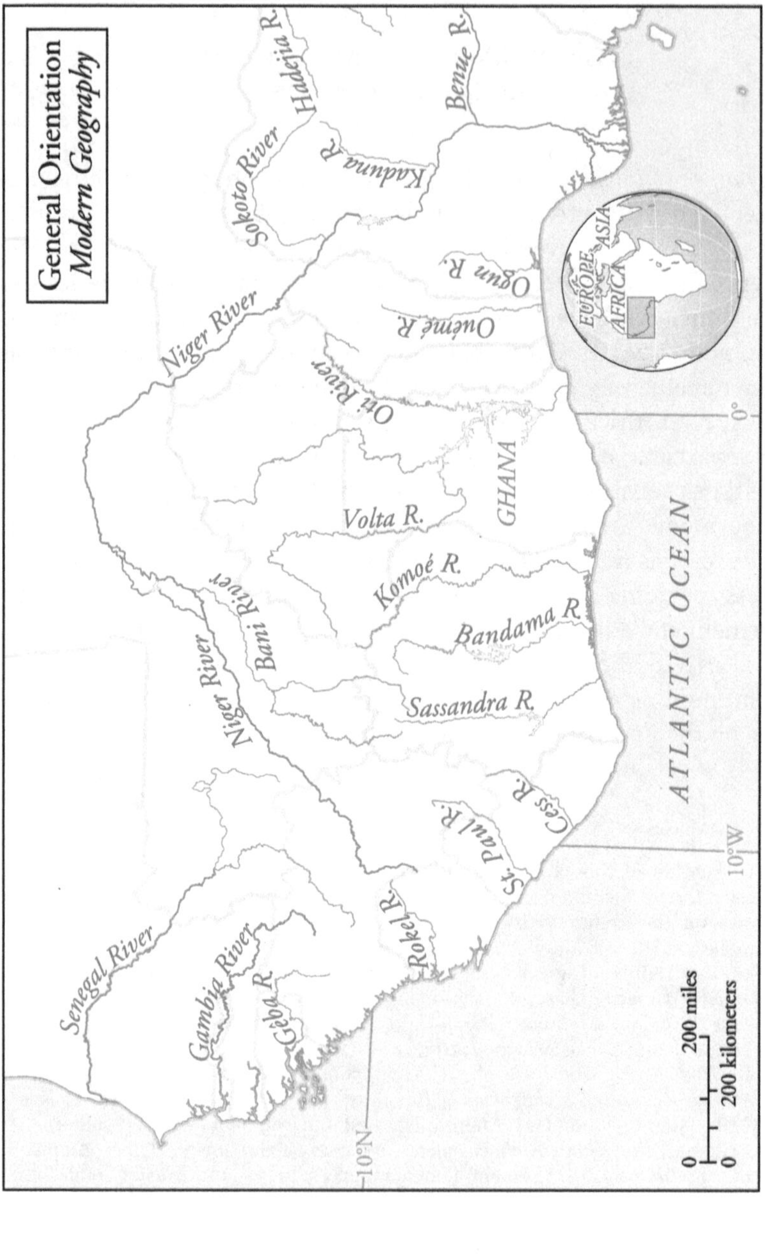


Map 1. General Orientation

### *Early Foraging to 25,000 BCE*

These foragers colonized grassland environments across the continent, including tropical West Africa during times when rainfall was less plentiful than it is now. They occupied areas that were then savannas, but which now are in deep forest [see Map 2].[4] The currently thick vegetation cover probably conceals most of the earliest settlement sites from easy archaeological discovery. But for the early foragers these historic grasslands were ideal habitats. Early foraging communities were small in size and scattered very lightly across the grasslands, but groups clustered on areas of particular food abundance. Like hunters and gatherers everywhere, they sustained themselves on wild plant and animal foods. Success with this food strategy rested on foragers' intimate knowledge of their environment, so that they could wrest good yields of the wild foods they preferred to eat, but with the least amount of effort necessary. Their mobility was a critical aspect of their strategy to sustain themselves with food. The foragers timed their movements to the changing seasons to collect wild foods as they came to ripen in different ecological niches over the course of a year. In the drier conditions of the deep past, foragers encountered the same sort of transition from increasingly dry and open country to the north and increasingly wet and wooded to the south.

Early foragers reconnoitered far and wide for habitats of particular food abundance to exploit and perhaps linger temporarily. They favored spots that offered potable water and rich ecological variation. To the north and east, the dominant region for repetitive clustering was alongside the Volta and its tributaries. South of the Kwahu escarpment and west of the Akuapem-Togo range, foragers favored rivers that ran not into the Volta, but directly towards the sea. Drinking water was their primary requirement, but these waters also supported freshwater fish and other riparian creatures to catch, and water-loving plants to eat. Areas of complex ecological variation offered a wide range of edible wildlife and year-round food, which reduced the foragers' movement and maximized food stability.

On both the open and wooded grasslands, the foragers' most important tool for procuring food was the controlled use of fire. The fires they set mimicked the effects of spontaneous flare-ups resulting from lightning

---

[4] M. A. Sowunmi, "The Environment: Present and Past of West Africa," *West African Journal of Archaeology* 17 (1987), 49; J. Maley, "Le cadre paléoenvironmental des refuges forestiers africains: Quelque données et hypothèses," in L. J. K. van den Maesen, *et al.*, eds., *The Biodiversity of African Plants* (Dordrecht: Klumer Academic, 1996), 519–35.

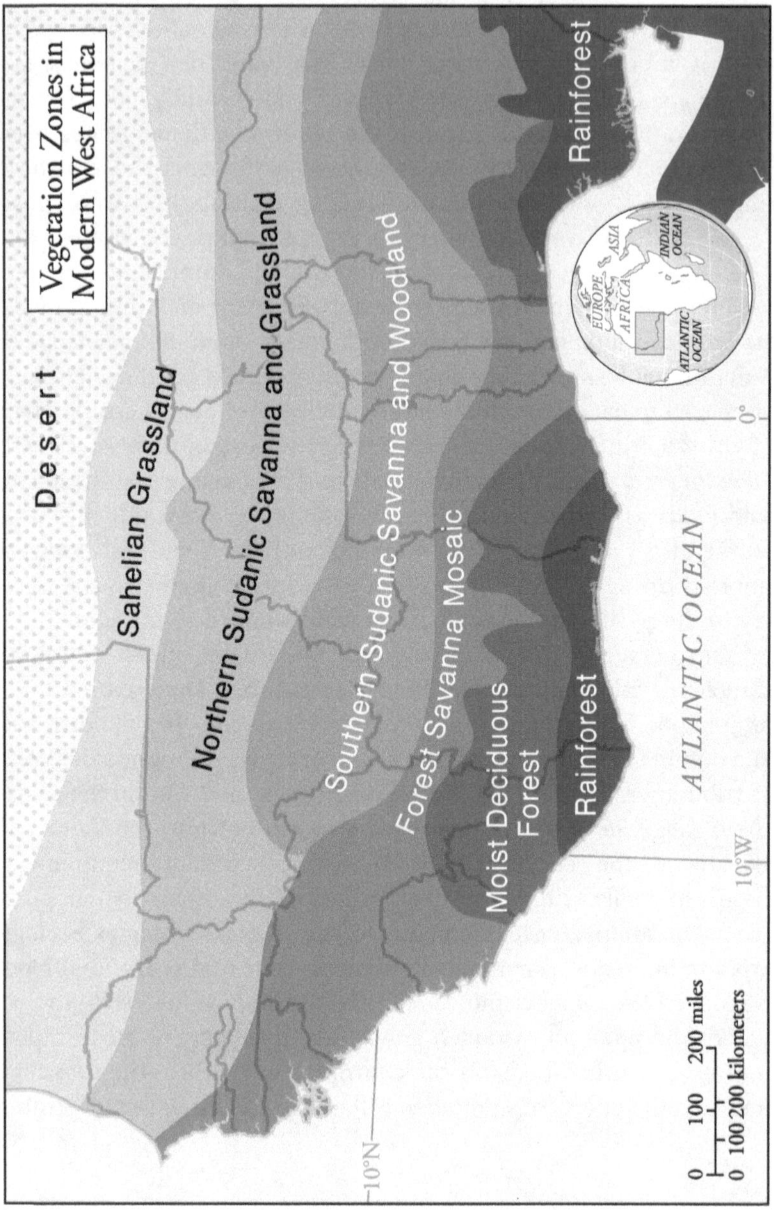


Map 2. Ecological Zones in Modern West Africa

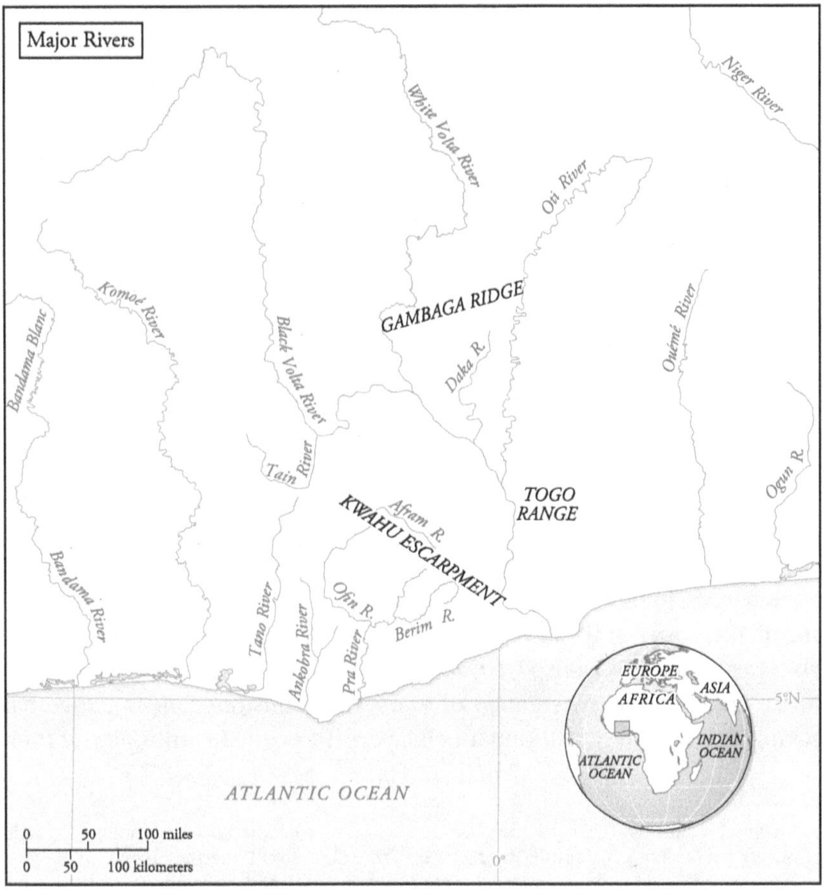


Map 3. Major River Systems of the Study Area

strikes or the hot fermentation of dead grasses, which sparked bush fires in the driest conditions.[5] They set fire to grasses thoroughly parched in the dry season to create pockets of hunting lands which attracted ruminant game when the rains returned and the grasses' underlying roots sent up the new, tender, and nutritious shoots their prey favored. But fire-setting had to be carefully timed and controlled, because some of the foragers' favored trees, including the Shea Butter tree (or the francophone karité, *Butyrospermum paradoxum* subsp. *parkii*), would not fruit if flames

[5] James Fairhead and Melissa Leach, *Misreading the African Landscape: Society and Ecology in a Forest-Savanna Mosaic* (Cambridge: Cambridge University Press, 1996), 46–47.

scarred the trees while they were flowering.[6] Foragers prized shea butter trees for the flavorful, fleshy fruit pulp which covered their nuts. Other trees, like the acacia and baobab, were resistant to burning and with time came to characterize the residual landscape of human habitation.[7] Varieties of acacia (*Acacia* spp.) attracted honey-making bees to their flowers, fattened ruminant game on their low-hanging leaves, and fed human foragers with their seeds.[8] Baobab trees (*Adansonia digitata*) dropped hardy brown fruits, which foragers collected and broke open to eat the citrusy-tasting pods inside.[9] Foragers plucked young baobab leaves to eat as a tender vegetable.[10]

By eliminating the tall growth of inedible grasses, fires also promoted the growth of one of the gatherers' most dependable foods, tubers tucked safely underground, which would otherwise have lost in the competition for light and moisture once the rains returned.[11] Important among these tuberous foods of these gatherers were West African yams (*Dioscorea* spp.), a classification that encompasses many modern varieties. *Dioscorea* had a tropical distribution worldwide even in pre-modern times, but certain species were indigenous only to the moister grasslands and forest margins of West Africa; these were the ancient, wild progenitors of the modern types of Yellow Yam (*D. cayenensis*) and White Yam (*D. rotundata*).[12] Direct archaeological evidence of early yam consumption is altogether lacking and probably will remain elusive.[13] To posit the antiquity of their

---

[6] Daniel K. Abbiw, *Useful Plants of Ghana*, (London: Intermediate Technology Publishing, 1990), 52, 66–67; D. Gledhill, *West African Trees* (London: Longman, 1972), 57–8.

[7] Harlan, de Wet, and Stemler, "Plant Domestication and Indigenous African Agriculture," 11–12; Gledhill, *West African Trees*, 54–55; Boateng, *Geography of Ghana*, 52.

[8] Abbiw, *Useful Plants of Ghana*, 13, 46, 53; Gledhill, *West African Trees*, 54–55.

[9] This tree's vernacular name, baobab, derives from the Arabic *bu hibab*, 'fruit with many seeds.'

[10] Gledhill, *West African Trees*, 54–55; Abbiw, *Useful Plants of Ghana*, 11, 41; Anquandah, *Rediscovering Ghana's Past*, 37.

[11] For the controlled use of fire in West Africa before the Holocene, see Coursey, "The Origin and Domestication of Yams in Africa," in Harlan, De Wet, and Stemler, eds., *The Origins of African Plant Domestication*, 400; for later eras, see Stahl, "Innovations, Diffusion, and Culture Contact," 55. For an extensive list of these tubers, see Abbiw, *Useful Plants of Ghana*, 30–1.

[12] For classification, see Coursey, "Yams in Africa"; and Bassey Andah, "Identifying the Early Farming Traditions in West Africa," in Thurston Shaw, Paul Sinclair, Bassey Andah, and Alex Opoko, eds., *The Archaeology of Africa: Food, Metal, and Towns* (London and New York: Routledge, 1993), 240–54.

[13] On the problems and prospects of finding evidence of early yams in this area, see Merrick Posnansky, "Early Agricultural Societies in Ghana," in J. Desmond Clark and Steven A. Brandt, eds., *From Hunters to Farmers: the Causes and Consequences of Food Production in Africa* (Berkeley: University of California Press, 1984), 151; Delores R. Piperno,

consumption argued here, historians must therefore rely upon agronomists' estimates that modern genetic varieties of West African yams represent human management reckoning back to the earliest human times in this region.[14]

For the early foragers, yams became an important wild food, but not a preferred one. They were important because they were plentiful, resistant to pestilence, and flourished in those grassland areas that foragers annually torched. But yams were also bulky, labor-intensive to dig and process, and foul tasting when they did. Unlike modern, cultivated varieties of yams, the early progenitors probably had very unsuitable qualities: they had prickly spines; the larger ones rooted especially deep and were thus difficult to unearth; and many early varieties contained varying amounts of toxins (*dioscorene*) that immediately discouraged foragers because of their strong bitter taste.[15] To process these bitter-tasting tubers into edible food, the foragers needed to open the root bulb (by slicing, shredding, or pulverizing), and then proceed to wash the starchy inner pulp in water in order to leech the toxins. The earliest yam gatherers had no pottery, but might have used water-filled calabashes or perhaps woven baskets immersed in running water to wash the tubers. Foragers may have added salt to the water to help leach the toxins; such salt might have come from coastal sources or inland mines, but probably was most often taken by rendering saline ashes from certain known plants that grew along river courses everywhere.[16] The detoxified cuttings were then probably processed into food much the same way as modern varieties, that is foragers applied heat and then pounded the wet starchy mass in a wooden pestle until they formed a soft, thick paste (like the modern *fufu*). Even with this extraordinary effort, yams were still just a bland bulk to fill foragers' bellies.

### *Specialized Foraging, 25,000 BCE to 10,000 BCE*

Succeeding intervals of environmental moisture and aridity after 25,000 BC stimulated foragers to reshape their food strategies to fundamentally

---

*Phytoliths: A Comprehensive Guide for Archaeologists and Paleoecologists* (Lantham MD: AltaMira Press, 2006), 48.

14 Coursey, "Yams in Africa," 400.

15 Coursey, "Yams in Africa," 395.

16 Abbiw, *Useful Plants of Ghana*, 30–31, 49.

changed habitats. Global climate changes shifted the West African environmental zones gradually northwards in wetter climate phases, and southwards in drier regimes. Desiccation rendered moist grasslands into dry scrub; increased rainfall changed dry woodland into thick rainforest. These environmental changes took place only very gradually,[17] but they had enormous repercussions on human communities. The bulk of foragers gradually drifted northwards and southwards with the environmental changes. Over the long term, their seasonal relocations to exploit familiar wild foods in familiar environmental zones created historic population flow.[18] But perhaps not all foragers retreated with their favored habitats.[19] Some may have chosen not to retreat northwards amid moister conditions, but rather to adapt their foodways to make a new living among the encroaching forest.[20] Similarly, in drying conditions, some foragers remained on desiccated environments. The necessity of remaining required them to invest more labor into finding food by expanding the geographic range of their foraging activities, so as to wrest adequate food from a resource-poor environment. Other foragers, residing in the most arid areas, abandoned mobility and clustered along the river courses where water was perennially available and still sustained some edible flora and fauna.

Historical climate change and subsequent human occupation of every environmental regime undergirded the growth of areally distinctive

---

[17] For detailed representations and a discussion of these pluvial changes, see A. Muzzolini, "Les climats au Sahara et sur ses bordures, du Pléistocène final l'aride actuel," *Empúries* 47 (1985), 8–27.

[18] Boahen, "Ghana before the Coming of Europeans," 96.

[19] Compare with the early settlement of Central Africa, in Jan Vansina, *Paths in the Rainforests* (Madison: University of Wisconsin Press, 1988), 47–48. For the applicability of Vansina's Central African models to early West African history, see Kevin C. MacDonald, "Korounkorokalé Revisited: the 'Pays mande' and the West African Microlithic Technocomplex," *African Archaeological Review* 14, 3 (1997), 195–96.

[20] Cf. the postulated settlement of the forest only after 9000 BCE, in Merrick Posnansky, "Archaeological and Linguistic Reconstruction in Ghana," in Christopher Ehret and Merrick Posnansky, eds., *The Archaeological and Linguistic Reconstruction of African History* (Berkeley: University of California Press, 1982), 259–60. Note that archaeological evidence for the forest settlement in the deep past is altogether absent. In West Africa, as in Central Africa, the earliest sites seem to relate to savanna environments; Vansina, *Paths in the Rainforests*, 47–49. But historical reconstructions describing forest settlement only in much later times seem to be based on this archaeological silence as much on the assumed inhospitable nature of the forest. Such assumptions of forest 'hostility' are common both to Western thought and modern metropolitan Ghanaian perception; see T. C. McCaskie, *State and Society in Pre-Colonial Asante* (Cambridge: Cambridge University Press, 1995), 9–10; and Vansina, *Paths in the Rainforests*, 43.

foodways that exploited the environmental niches of their wild plant and animal foods. Savanna foragers specialized in the collection of fruits from baobab and shea butter trees. They also collected the edible underground parts of some plants (i.e., roots, rhizomes, corms, and tubers) and the seedy tops of others. They supplemented these plant foods with the hunting of ruminant game, birds, and small rodents. To the south, in the dry forest, foragers concentrated their efforts on collecting tree fruits from oil palm (*Elaeis guineensis*) and hackberry (*Celtis* spp.) trees, some palatable leaves, and snails. They hunted and trapped native creatures like bush pigs, rodents, and small primates. Rivers everywhere sustained good food to collect, like freshwater shellfish, frogs, and, most importantly, fish.

The foragers' food strategies, despite local and historical articulation, were broadly similar. They gathered wild plant foods that required only basic tools to reap and which needed little preparation or processing thereafter to turn into food. Those who knew the territory well enough foraged for eggs, insects, and snails, and they hunted and trapped small game and fish; these activities together probably brought more protein and total calories into their diets than heroic hunting of dangerous large animals did. The romantic historiographical stereotype of hunter-gatherers' innocent state of social equality and economic communalism is probably too simplistic and is certainly premature based on the evidence at hand. Only future archaeological research will illuminate the social composition and ideologies of these early people. But if these settlers were, again, like modern hunters and gatherers elsewhere in Africa, then their sharing of food might have been selective and strategic. Some might have shared food to establish and perpetuate social relationships, or to temper social discord.[21] Others might have shared windfalls of perishable food that would otherwise spoil or to reward partakers in successful cooperative ventures.[22]

To 10,000 BCE, early inhabitants lived in small communities and understood their social relationships, it is presumed, primarily in terms of familial relatedness. These communities were very small in size and the total

[21] O. Gulbrandsen, "On the Problem of Egalitarianism: The Kalahari San in Transition," *Journal of Comparative Social Religion* 10–11 (1991), 12–31.

[22] Referring specifically to hunted meat, but suggestive for other foods, including plants, see R. Layton, R. Foley, and E. Williams, "The Transition between Hunting and Gathering and the Specialized Husbandry of Resources: a Socio-Ecological Approach," *Current Anthropology* 32 (1991), 258.

population of the area was minuscule compared to modern numbers.[23] These scattered small communities were highly mobile and organized their movements and their temporal and cultural calendars around the annual cycle of changing seasons. These foragers expertly exploited their environment to sustain themselves with wild foods. Despite their sparse numbers and independent subsistence economies, foraging groups did not live in complete social isolation. Seasonal patterns of residential mobility increasingly brought groups into contact as they moved to exploit diminishing geographically specific natural resources. Such periodic contacts would have been of substantial importance if they included, through marriages, biological exchanges and cultural contacts.[24]

### *Intensified Foraging from 10,000 BCE*

In the years after 10,000 BCE foragers began to integrate food production into their strategies of subsistence upon wild foods. They began experiments that entailed developing closer relationships with what they ate. They started to invest more time and attention into reaping food from wild sources. In doing so, they blurred modern typological distinctions between foraging and farming strategies, and between wild and cultivated foods. In this era, foragers also sampled new wild foods and they investigated novel ways to make meals by fashioning new tools and techniques. They also experimented with becoming more sedentary and accumulating more material things.

Foragers did not realize these changes as a sudden sweep of revolutionary progress. Archaeological evidence for this period and area correlates very poorly with the evolutionary notion that the older, autochthonous, mobile hunter-gathering population was replaced at the end of this era by a new, invasive, sedentary, farming people who migrated from the Sahara southwards with superior technologies, including, most importantly, pottery and iron. These intrusive groups were normally described as 'Neolithic,' a term borrowed from European archaeology and married

---

[23] The only attempt I have seen to calculate the population prior to the Atlantic era is in Posnansky, "Archaeological and Linguistic Reconstruction in Ghana," 264. Posnansky's rough calculations pointed to fewer than 50,000 persons in 1,000 CE—which I use to calculate a density of approximately one person per five square kilometers (when using the area of the current state of Ghana).

[24] The general economic condition of subsistence foragers is described in Ralph Austen, *African Economic History* (London: James Currey / Portsmouth NH: Heinemann, 1987), 9–29.

to the notion that agriculture was such a difficult idea to invent that only a handful of human societies, notably those of the Fertile Crescent and Sahelian West Africa, initiated an 'agricultural revolution' that diffused outward from those singular centers as an obviously superior way of living. The continued resiliency of this diffusionist and evolutionary model in West African historical studies owes much to the tremendous difficulty of attributing sequence and motivations to foragers' early experiments at farming and sedentarism. The picture that emerges from recent interdisciplinary research hints to a much more complex and historical past than the evolutionary model can accommodate. It now seems likely that the 'agricultural revolution' was rather a very slow, incremental transformation of a great many existing techniques into an unanticipated transformation. Over the course of several millennia, foragers intensified their collecting strategies, going slowly and even abandoning laborious innovations.

Foragers began their experiments in the context of a long-term climatic phase, known as the Holocene (beginning 10,000 BC) and which brought increased moisture and temperature to their habitats. The Holocene is our current geoclimatic era; its onset brought to West Africa ecological conditions and annual seasonal patterns broadly similar to those of today. Despite increasingly sophisticated climate histories describing pluvial changes within the Holocene era, it remains terribly difficult to infer specific vegetational changes in local habitats with similar precision. Nor does it seem that foragers' experiments were just a simple reflexive reaction to environmental change. Surmising the significance, if any, of specific climatic events and their inferred environmental impacts on the course of human history thus continues to offer ample opportunity for risky speculation. While there is no doubt that changes in the foragers' physical environments figured in their decisions to experiment with new ways of living, foragers' efforts sprang equally from their own curiosity and creativity.

The onset of these long-term, wet conditions was very gradual. As in previous climactic phases, many foragers maintained their preference for forest-savanna mosaics, and they followed those favored niches as they retreated northward. Others remained to reshape their foodways to moister niches.[25] But this period, unlike previous eras, was punctuated by more frequent and more severe short-term oscillations between aridity

[25] S. K. McIntosh and R. J. McIntosh, "West African Prehistory (from 10,000 BC to AD 1,000)," *American Scientist* 69 (1981), 602–13; Boahen, "Ghana Before the Coming of Europeans," 96.

and abundant moisture.[26] Such abrupt climactic events confronted foragers with extreme unpredictability. To the area's foragers, these short-term events were serious and utterly unpredictable. While short-term climactic events left little permanent imprint on local environments, they interfered sufficiently with the annual supplies of foragers' wild foods on which they depended. Amid drought, foragers' favored trees refused to flower or they would drop their fruits early before they had a chance to mature into something edible. Extraordinarily wet conditions would foster weeds that competed with favored food plants, and encouraged the growth of fungi, which spoiled fruits and rotted tubers sitting in waterlogged soils.

Foragers in various environments perceived unpredictability in different ways, and thus the activity and the intensity of their early experimental efforts to handle instability varied areally and historically. Foragers aimed early experiments at farming towards perfecting their foraging strategies. Foragers simply protected and nurtured their sources of wild foods, spending a bit more effort to insure that their wild sources of foods flourished every year. Such tending of wild food sources was probably ancient rather than innovative, but in this era foragers increased their efforts to promote good growth, fruiting, and reproduction.

By tending and managing these sources of food, foragers altered the environments in which they grew, influencing the distribution and characteristics of wild plants to degrees that altered other aspects of their lives. This relationship was symbiotic; foragers also began to reshape their tools, techniques, and strategies to adapt to the changes they achieved in these wild foods.[27] What foragers began in these very early times was the *process* of domestication. Domestication was not an event, nor a self-evident genetic result. Rather, domestication was a process without a conclusion, because the relationship between humans and their foods is historical. That is, it changes over time. Descriptions of 'domesticated' crops and animals therefore demand historical specificity: who adapted them?; under what circumstances?; what characteristics were sought?; and what were the results, unintended as well as intended? The characteristics of these adapted plants and creatures never remain static, because people—then

[26] Muzzolini, "Les climats." The theme of unpredictability is developed in Roderick J. McIntosh, "The Pulse Model: Genesis and Accommodation of Specialization in the Middle Niger," *Journal of African History* 34, 2 (1993), 194–205.

[27] This processual relationship is described as 'co-evolution' in D. Rindos, *The Origins of Agriculture* (Orlando: Academic Press, 1984).

as much as now—constantly try to adapt their foods to changing environments and their own, however contested, tastes and preferences.[28]

Foragers on open northern grasslands sensed the greatest vulnerability to recurrent shortfalls in wild foods. On grasslands at the headwaters of the Volta, the edible flora they preferred to eat was highly dependent on the area receiving adequate rains during a brief mid-year wet season. But these northern areas were open grasslands because the rains there were very fickle, and foragers saw many lean years when their wild foods were scarce.[29] Grassland foragers targeted their efforts at enhancing returns on the most stable resources that were diminished least amid highly variable weather. Grassland foragers insured that young specimens of their favored tress were not scorched by their annual bushfires. They spared valued resources by cutting firebreaks, and setting so-called 'early' fires at the beginning of the dry season, when morning dew and residual plant moisture retarded flames from spreading, so as to consume fuel close to resources.[30] They also cleared, with a bit of muscle, weedy growth from around these favored plants, including young baobab, acacia, hackberry, and shea butter trees, so that they might prosper without substantial competition for sunlight and ground moisture. Foragers especially prized the oily seeds from canarium (*Canarium schweinfurthii*). Women collected canarium bark resin to mix with shea butter in order to make a fragrant skin lotion. The tree's modern vernacular name speaks to the perfume's ability to attract men (Akan, *bediwouna*, 'you would have sexual relations with your sister').[31] Foragers continued to protect young trees until they were tall enough to no longer be shaded by grasses, and large enough to withstand annual fires without human help.

Along the length of the Volta in this era, foragers intensified their exploitation of riparian resources. They clustered their settlements alongside the upper reaches of the Volta river, where they lessened their seasonal mobility in favor of spending long periods along the riverbanks. In essence, they converted what was once a dry-season activity into a more

---

28 Identifying desirable characteristics is never done in a purely neutral, technical manner. The desirability of new domesticates is always contested within communities, as is obvious in the contemporary rancor over 'genetically modified' crops.

29 E. Ofori-Sarpong, "Rainfall Reliability over Ghana," *Journal of Eastern African Research and Development* 21 (1991), 93–106; Boateng, *Geography of Ghana*, 22–37, 74–5; J. Papadokis, *Crop Ecologic Survey in West Africa* (Rome: FAO/UN), maps 4, 8.

30 Fairhead and Leach, *Misreading the African Landscape*, 230–33.

31 Albert A. Enti, *The Rejuvenating Plants of Tropical Africa* (San Francisco: North Scale Institute, 1988), 25–26.

permanent subsistence base. Still, some river fisherman in the far north had to leave the riverbanks at the very height of the dry season when the rivers might dry up completely. Foragers improved their technology and began to fish with barbed ivory hooks, and probably also with nets and plant-derived stun-poisons. Foragers also preyed upon other water-loving creatures, such as turtles, hippos, and freshwater shellfish.

More than in any other niche, foragers downstream at the mouth of the Volta fashioned their subsistence strategy on intensive exploitation of watery life. These coastal areas remained very dry, even with the onset of the moister conditions of the Holocence. Annual rainfall patterns that brought more rain to the south than the north were here significantly disrupted by the upswell of a cold ocean current. Moreover, what rain did fall was quickly evacuated by porous soils. Foragers there inhabited an environment that more closely resembled the far northern grasslands than they did nearby forest niches.[32] Moreover, on these grasslands at the mouth of the Volta, rainy seasons were even briefer than on the northern savannas, and cumulative annual rainfall amounts were even more variable.[33] Foragers here fished year-round, using combinations of dugouts, nets, and fishing weirs to snare different species of fish and shellfish from fresh and brackish waters.[34] Coastal fisherfolk were perhaps able to apply some of their river and lagoon fishing techniques to the open sea.[35] Thereafter, fisherman settled ideal spots along the entire coast, from the dry grasslands at the Volta mouth to the forested western coast, where they could integrate exploitation of river, lagoon, and saltwater fisheries.[36]

---

[32] Nygaard and Talbolt, "Stone Age Archaeology and Environment on the Southern Accra Plains," 20; J. Jeník and J. B. Hall, "Plant Communities of the Accra Plains, Ghana," *Folia Geobotanica et Phytotaxonomica* 11 (1976), 163–212; Stahl, "Holocene Archeology of Ghana," 56; Boateng, "Geography of Ghana," 52–53.

[33] E. Ofori-Sarpong, "Rainfall Reliability over Ghana"; Boateng, *Geography of Ghana*; H. P. Walker, "Weather and Climate," in J. B. Willis, ed., *Agriculture and Land Use in Ghana* (London: Oxford University Press, 1962), 7–50; Papadokis, *Crop Ecologic Survey*, maps 4, 5, 6, 8.

[34] Andah, "Asokrochona Coastal Region"; Nygaard and Talbolt, "Stone Age Archaeology and Environment on the Southern Accra Plains."

[35] The notion of indigenous ocean travel is controversial. Boahen dated early oceanic voyaging to before Portuguese contact with the Gold Coast; Boahen, "Ghana before the Coming of the Europeans," 99–104. Arguments against West African seafaring point to insufficient indigenous boat technology; for example, see Boateng, *Geography of Ghana*, 111.

[36] Nygaard and Talbolt, "Stone Age Archaeology and Environment on the Southern Accra Plains"; Robert Chenorkian, "Ivory Coast Prehistory: Recent Developments," *African Archaeological Review* 1 (1983), 127–42.

In the rainforests in the hinterland of the coast, foragers perceived little insecurity. There the rains fell more consistently throughout the year and annual fluctuations were muted. After 10,000 BCE the population of the rainforest remained very small relative to other habitats, and forest-dwellers were scattered areally in spots of particular wild food abundance.[37] They collected nuts from the hackberry, canarium, and oil palm trees. Forest foragers' favored trees also attracted game, like colobus monkeys (*Colobus* spp.) and bush pigs, which competed with foragers for foods still on the tree, or that had already fallen to the ground, respectively. Forest foragers adapted their lithic technology towards smaller, lighter stone points suitable for fixing onto projectiles. With the adoption of the bow,[38] hunters were able to reach the forest canopy where most of the niche's birds and primates resided. Nets and snare traps set at forest floor caught other sorts of animals, like bush pigs, guinea fowl (*Numida* spp.), forest duiker (*Cephalophus* spp.) and various rodents.[39]

### *From Foraging to Farming*

After 10,000 BCE foragers on the northern face of the Kwahu Escarpment began consistent and significant efforts at landscape management. There they found their ideal ecotone habitats slowly encroached by forests expanding with the moisture that this new climate phase brought north. To maintain the open character of the ecotones, foragers there experimented with retarding high forest growth so that their favored wild trees—canarium, oil palm, and fan palm (or borassus palm, *Borassus aethiopum*)—might have adequate sunlight to flourish.[40] Wild yams, too, would not grow without adequate light. Maintaining clearings was done with stone axes used to keep check on spontaneous forb regrowth.[41] Foragers left cut brush to dry until the end of the dry season, at which time they would set it alight. After the brush burned, foragers spread the

---

37 Posnansky, "Archaeological and Linguistic Reconstruction in Ghana," 259–60.

38 Anquandah, *Rediscovering Ghana's Past*, 54.

39 Referring to the end of this era, see Stahl, "Holocene Archaeology," 76.

40 Posnansky, "Archaeological and Linguistic Reconstruction in Ghana," 259–60; Stahl, "Holocene Archaeology," 71; S. R. J. Woodell and L. E. Newton, "A Standard-Winged Night-Jar Breeding in the Forest Zone of Ghana," *Nigeria Field* 40 (1975), 169–71.

41 Coursey described the utility of stone tools for clearing in D. G. Coursey, "The Origin and Development of Agriculture in Africa," in Harlan, De Wet, and Stemler, eds., *The Origins of African Plant Domestication*, 397.

ashes over the clearings to improve soil quality. These open areas were small and dispersed.[42]

Foragers of these maintained ecotones and those of other natural openings experimented in this era with managing wild yams. Yams were presumably still something of a famine food rather than a preferred food but in a pinch could be turned into stomach-filling staple. Foragers used social controls to maximize yam resources in case they might be needed in this era of unpredictability. They generated communal prohibitions against harvesting young, or 'new,' yams by individuals until group leaders gave their consent. Authority in such regards was not arbitrary but reflected the foragers' hard-won knowledge about the tuber. The prohibition against 'early' harvesting corresponded to the phase of most active growth of the yam plant. Prematurely unearthed yams would give only a very small yield of tubers and with still-soft skins that were highly susceptible to damage and thus spoilage.[43] For the experimenters, managed yams were a fail-safe food source. If the wild foods they preferred were available and abundant, then they would leave managed yams in the ground to reproduce and, perhaps, be harvested and eaten the following year, as needed.

Ecotone foragers also experimented applying sophisticated vegeculture techniques on wild yams. Foragers identified desirable characteristics, such as large tubers and a 'sweet' taste indicative of low *dioscorene* toxicity, in certain wild yams and encouraged their reproduction. Perhaps those responsible for the arduous processing of mildly poisonous wild yams into famine food—proposed here as women—might have been those most keenly interested and involved in adapting the wild progenitors into less poisonous, cultivated types. Experimenters intervened in the reproductive cycles of their trial crops to gradually produce varieties with qualities that these innovative foragers favored. Some ancient types of wild yams probably reproduced sexually to produce hybrid seeds, but experimenters could also reproduce preferred yam specimens by taking cuttings of them. If planted in ground with sufficient moisture and soil nutrients (including compost piles), the nitrophilic yam heads would produce genetic clones of the parent stock.[44] Yam genes were very plastic however, and so it was

---

[42] Merrick Posnansky, "Early Agricultural Societies in Ghana," in J. Desmond Clark and Steven A. Brandt, eds., *From Hunters to Farmers: the Causes and Consequences of Food Production in Africa* (Berkeley: University of California Press, 1984), 149–50.

[43] Cf. the proposition that these communal controls date from before 13,000 BCE; in Coursey, "Yams in Africa," 398–99, 400.

[44] Coursey, "Yams in Africa," 400–1; Andah, "Identifying the Early Farming Traditions of West Africa," 240–54. I am suggesting a conscious effort to adapt wild African yams. Other

not rare that experimenters found that the early crops they planted had spontaneously mutated into new varieties with surprising, or disappointing, characteristics.[45] Through testing, cutting, and replanting human selection produced the genetic types of the modern cultivars. Forest clearings and boundary areas between wooded country and grasslands were the most productive areas for experimentation, as they offered substantial sunlight, good moisture, and a range of micro-habitats in which new trial crops could be tested and observed closely.[46] Experimenters also produced types better suited to local environmental conditions, including varieties suitable for drier conditions (White Yam, *D. rotundata*) as well as more humid niches (Yellow Yam, *D. cayensis*).[47]

Early experimenters soon learned that their gardens needed to be protected from wildlife pests. Monkeys, bush pigs (*Potamochoerus* spp.), grasscutters (or cane rat, *Thryonomys swinderianus*), and probably elephants, too, preyed upon the fruits of farmers' labors. Some garden experimenters planted varieties of wild yams with high toxicity (such as *D. dumetorum*, which induced convulsions when consumed) to encircle their food crops and thereby deter monkeys and other thieves.[48] Other planters enclosed their gardens with fences, and positioned traps at key points to snare bush pigs and grasscutters. Foragers' expertise with traps protected their crops at the same time they snared valuable meat to consume. The clearings also attracted other woodland game, like forest duikers and tortoises, which were not garden pests but were likewise attracted to open areas in the forest. Trapping and garden hunting alleviated the need for hunters to spend extended periods of time away from the village to search for game.[49] Hunters stayed busy however, trying to kill garden predators that traps could not catch. Grasscutters were notoriously savvy

seedy crops may have been developed without such intent; the 'rubbish heap' model is cogently described in Jared Diamond, *Guns, Germs, and Steel: the Fates of Human Societies* (New York: Norton, 1997), 114.

[45] Coursey, "Yams in Africa," 400n5, 401.

[46] Harlan, *et al.*, "Plant Domestication and Indigenous African Agriculture," 18; Coursey, "Yams in Africa"; Bassey Andah, "Agricultural Beginning and Early Farming Communities in West and Central Africa," *West African Journal of Archaeology* 17 (1987), 180.

[47] Coursey, "Yams in Africa," 400–1.

[48] Andah, "Identifying the Early Farming Traditions in West Africa"; Abbiw, *Useful Plants of Ghana*, 27, 210.

[49] Ann Stahl, "Reinvestigation of Kintampo 6 Rock Shelter, Ghana: Implications for the Nature of Culture Change," *African Archaeological Review* 3 (1985), 143–44; Jan Vansina noted similar importance of garden trapping in the foodways of Central Africa; Vansina, *Paths in the Rainforests*, 90.

at foiling traps and always needed to be hunted if gardeners were to protect their crops. No trap or fence would stop hungry elephants, and thus they were always feared for the damage they could wreak on gardens, homes and, even, people.

Ecotone inhabitants found protected and cultivated yams to be reliable insurance against failures of less attended sources of wild food. But this insurance came at a moderate cost. Foragers learned that they could not continually tend yams on the same cleared spots year after year without suffering dramatically diminished returns; these 'improved' varieties of yams required rich soils. They did their best to give them that, using stone blades like hoes to scrape up the nutrient-rich top layer of dark topsoil into mounds into which they planted the yams' heads.[50] Mounds of rich earth provided, in the short term, abundant nutrients and allowed cultivated yams to grow larger in loose earth than those wild types could ever achieve in compacted, uncultivated ground. Mounds also lifted cultivated yams above waterlogged soils in the rainiest months of the growing season, and thus protected the tubers from disease and rot. At the end of the growth cycle foragers found yams grown in mounds to be easier to uproot and harvest, with a simple digging stick, than those in fixed earth.

At the end of this annual calendar, used soils were no longer rich enough to support another crop of yams. Experimenters used these old gardens for trials with other food plants, including various sorts of greens (or, 'spinach'), pulses, and squash,[51] as well as to tend plants with medicinal uses[52] or craft importance, such as bottle gourds.[53] But for new yams, virgin patches of ground had to be cleared if the ennobled types were to flourish. Foragers searched in their local vicinities for suitable locations for new plots. They sought natural openings that alleviated the heaviest clearing work and they paid attention to certain indicator plants, like the

---

[50] These grounded-edge stone tools are know in archaeological literature as *nyame akuma* (Akan, 'god's axe'). For a description of these tools and the suggestion that stone hoes continued to be used even after the adoption of iron tools, see R. S. Rattray, *Ashanti* (Oxford: Oxford University Press, 1923), 321–34; Cf. Posnansky, "Early Agricultural Societies in Ghana," 150. Coursey put the development of specialized tools for yam culture at 45,000 to 15,000 years before present; Coursey, "Origin and Domestication of Yams in Africa," 400.

[51] Abbiw, *Useful Plants of Ghana*, 31–43.

[52] Useful list of medicinal plants, and descriptions of their specific efficacy are in Abbiw, *Useful Plants of Ghana*, 118–205; E. Ayiteh-Smith, *Prospects and Scope of Plant Medicine in Heath Care* (Accra: Ghana Universities Press, 1989); Enti, *The Rejuvenating Plants of Tropical Africa.*

[53] Abbiw, *Useful Plants of Ghana*, 103–17.

silk cotton tree (or kapok, *Ceiba guineensis*, which was also the source of medicines, craft materials, and edible leaves) that thrived on the richest soils.[54] Ideal spots were those nearest to the their villages, which minimized the need to walk too far to fields and to haul harvested foodstuffs back to their settlements. Foragers accustomed to mobility but intensifying their devotion to incipient cultigens saw no contradiction in moving their home bases—almost villages—to new areas when the exploited soils near where they were living gave out. In other cases, scarcities at home must have motivated younger men to set off by themselves to exploit terrain still left in its natural state.

In this way, communities spread their experiments with rudimentary agriculture broadly along the grassy margins of the forest, perhaps even more rapidly than similarly small populations entirely reliant on established agriculture would have dared. There, experimenters enjoyed adequate moisture and relatively good soils in which to plant their trial crops. In these areas, experimenters were many. Although all experimenters focused their efforts at producing a reliable food resource with good taste and although they might have had limited contact with one another, experimenters conducted their trials independently. In this sense, there was no single center of 'domestication' and agricultural revolution in West Africa from which agricultural strategies subsequently and quickly emanated.[55]

Experimenters applied their creative skills to food crops for several millennia after 10,000 BCE. During this era of experimentation with what became early farming, these foragers continued to rely primarily on wild foods they collected. Experimenters farmed improved types of yams at the same time they continued to collect suitable wild types. While these foragers placed cultivated yams in the significant role of insurance food, they continued to value certain types of wild yams.[56] Although they were more sedentary than foragers who did not tend gardens, they did not abandon strategic mobility to exploit wild foods over the annual changing of

[54] Akwasi Boakye-Boaten, "Cultural Ecology of Asante, 1702–1945" (unpublished PhD diss., University of Ghana, Legon, 1974), 139. Boaten noted that *C. pentandra* was the indicator for Asante farmers, presumably in more recent times. This particular species is a colonial era hybrid; see Gledhill, *West African Trees*, 18. The ancient, native species is *C. guineensis*; see Hepper, *West African Herbaria of Isert and Thonning*, 31. On the uses of various parts of the tree, see Abbiw, *Useful Plants of Ghana*.

[55] Coursey, "Yams in Africa," 395–96.

[56] Abbiw, *Useful Plants of Ghana*, 27; Coursey, "Yams in Africa," 390–401.

seasons.[57] Amid their travels they surveyed new areas with good potential to make into successful gardens and they noticed wild foods that they could potentially adapt through experimentation into new garden crops. Experimenters' mobility supported the geographic expansion of garden farming. Initially they sought plots within the moist woody borderlands but thereafter they adapted their food production techniques to neighboring niches.

Foragers of the ecotones began to explore choice parts of the semi-deciduous forest to the south and dry savanna-woodland to the north after 3000 BCE [see Map 4]. They initially colonized micro-environments which abounded with familiar wild foods to collect and also had prime garden spots, with adequate moisture and good soils, where they could plant their core crops, choosing spots where they found wild varieties of their core crops. Wild stores were initially only foraged. Later, these spots might be turned into gardens for cultivated types. Their colonization of these rich spots was not unproblematic. As they moved away from the heartland of experimentation in the ecotones, foragers had to adapt the plants they managed to the significant challenges posed by subtle, areally specific differences in seasonality, moisture, and soil profiles presented to their efforts to introduce farming methods developed in other conditions. Thus as the frontier of experimentation expanded outward, the era of experimentation began anew as foragers sought to apply the techniques of ecotone agriculture to new niches.

The cultivating foragers of the northern savannas were attracted primarily to the forestlands lying south of the ecotones.[58] There, they found garden spots with rich soils, which they noted by the topsoils' dark brownish color[59] and by the growth of certain indicator plants that grew on nutrient-rich earth. Settling foragers also found that the rains fell in forested country more abundantly and with greater perennial reliability than on the savanna woodlands. They first exploited the natural openings in the forest, which were similar to their familiar ecotone environments.[60]

---

[57] Seasonal mobility is stressed in Stahl, "Holocene Archaeology of Ghana," 71–72, 78.

[58] Posnansky dates agricultural settlement of the forest to before 1,000 BC; Merrick Posnansky, "Archaeology, Technology, and Akan Civilization," *Journal of African Studies* 2, 1 (1975), 26.

[59] In general, dark soils in the forest were rich in nutrients, and pale soils were unsuitable for agriculture; Boateng, *Geography of Ghana*, 57.

[60] The earliest forest sites seem to correlate with natural savanna inliers; see for example, L. E. Newton and S. R. J. Woodell, "A Newly Discovered Site for the Kintampo 'Neolithic' Cultural Tradition," *Sankofa* 2 (1976), 19–22; James Anquandah, "Boyasi Hill: A Kintampo

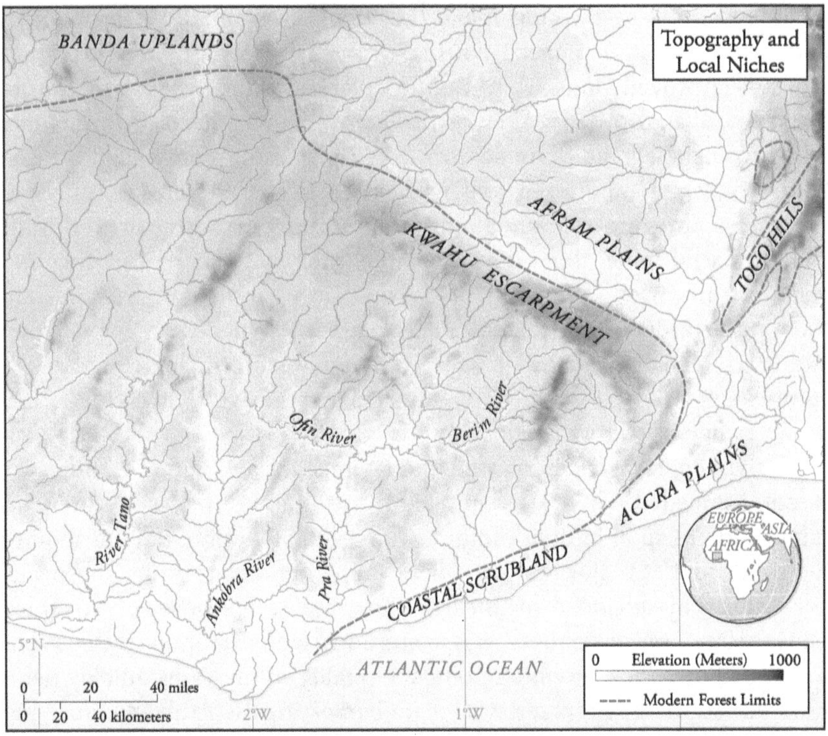


Map 4. Topography and Local Niches of the Study Area

There they applied familiar technologies and crops adapted to those conditions. At the appearance of the dry season, when the silk cotton and other deciduous trees began their annual shedding of leaves, cultivating foragers began to prepare gardens. They used sharp-edged stone axes to cut and clear light undergrowth. They left large trees standing to provide the dappled shade that nurtured young crops. Preparers then left this brush to dry to the end of the dry season, when they returned to burn cut growth was and spread the resulting ashes over the plot as a fertilizer.[61] They then scraped together the topsoils into mounds, removing

Culture Site," *Sankofa* 2 (1976), 92. Ivor Wilks observed that the savanna characteristic of these sites in modern times might be misleading as to their earlier 'natural' forested state, and that these clearings might have been of human making in ancient times; Ivor Wilks, "The State of the Akan and the Akan States: A Discursion," *Cahiers d'études africaines* 22, 3–4 (nos. 87–88) (1982), 236, 236n3.

[61] Clearing techniques falling short of clear-cutting are quickly summarized in A. Norman Klein, "Slavery and Akan Origins?," *Ethnohistory* 41, 4 (1994), 647n5. A description of

rocks and roots that would make the tubers difficult to unearth intact and unscarred, and planted cuttings from a variety of their preferred types of improved yams. They stopped planting yams when a migratory butterfly (now known in Akan as *akobonso*) showed itself in immediate advance of the great rainy season.[62] With planting completed, garden cultivators then monitored and weeded their garden yams throughout the rains. Foraging remained important during this season and they collected wild foods that came available, such as giant snails (*Achatina achatina*) that were abundant in the first six weeks, in order to sustain them until garden produce could be harvested.[63] When the rains stopped and it was decided that yams were completely mature (some signs were completely visible, for example by checking the withering vine), they began to harvest their yams.

Cultivating foragers harvested their yams throughout the short dry season as needed. There was no need to rush harvesting, as mature yam tubers did not spoil in their mounds. Harvesting therefore did not require the same sort of major labor input as preparing and planting these gardens. But they did need to unearth yams before the onset of the short rainy season, because these wet conditions would stimulate the tuber out of its dormancy, at which point the quality of the yams quickly deteriorated. Cultivating foragers took care harvesting their tubers before the rains returned because hasty or rough handling would damage the tubers' dark skin and thus invite spoilage of its inner, edible part. Properly handled, cultivators found that harvested yams stored well, if they were kept dry. In a few restricted areas where the short rainy season was reliable and would last for 90 days, yam farmers may have decided to leave some unearthed yams in their mounds to go through another growth cycle.[64]

While garden planters were attracted to the moisture that fell on more heavily forested spots further and further south and southeast,

---

twentieth-century garden preparation, including burning rather than felling large trees, is excerpted in Wilks, *Forest of Gold*, 45.

[62] Boaten, "Cultural Ecology of Asante," 138–9; Nana Akwasi Abayie Boaten, "Asante: The Perception and the Utilization of the Environment before the Twentieth Century," *Research Review* 6, 2 (1990), 26–27.

[63] On the importance of snail collecting, see McCaskie, *State and Society in Pre-Colonial Asante*, 29.

[64] This 'second season' yam strategy has been obscured in modern agricultural calendars, because farmers now choose to sow various sorts of introduced starches in this season. There is a faint record of it in Anne Klingelhoffer, "Agriculture," in David Brokensha, ed., *Akwapim Handbook* (Tema: Ghana Publishing Corporation, 1972), 134.

making gardens there required greater amounts of labor.[65] In this direction, cultivating foragers found the dry season increasingly short. As their agricultural labors were primarily concentrated in the dry season—with cutting, clearing, and planting—people who desired to make gardens in these forests needed to collect ample labor.[66] The brevity of the dry season also made burning brush problematic, because growth cut at the beginning of the dry season might not be dry enough to burn before the rains returned, when cultivators needed to prepare yam mounds and begin planting. In that case, the hard work of carrying away brush by hand, rather than burning it in place, was necessary. Also, they found that, in general, soils in this direction were less fertile and that the rich layer of dark topsoil became increasingly thin. Forested lands were typically hillier and therefore once gardens had been cleared of brush, the denuded slopes were prone to erosion when the rains began to fall. The fragility and low fertility of the soils made the arduous creation of new gardens in the forest an annual necessity in the forest.

Garden makers found the evergreen rainforest lying further southeast, beyond the densest deciduous woods, to be quite unattractive to agricultural settlement. Only extreme amounts of labor input were sufficient to carve gardens out of these forests. Rains fell there nearly year-round, vegetation to be cleared was denser, soils were even poorer, and hillier terrain made erosion unavoidable. The improved varieties of yams were adapted to a dry season; indeed, the purpose of this plant's tuber in its wild, perennial state is to retain water and store nourishment to sustain the plant through dry season dormancy and then support new vine growth when the rains return. In this direction, the appearance and duration of the dry season was most unpredictable. Without a distinct dry season, yam crops would at best fail to prosper or at worst completely fail to produce tubers. Therefore the wettest forest regions remained lightly populated compared to the moister savannas and deciduous forests. In the coastal areas roughly between Lagune Ndogo and the Pra River, communities maintained their primary dependence on wild fruits and game.[67]

---

65 The awesome labor input needed to completely clear forest gardens of tree growth is cogently described in Wilks, *Forests of Gold*, 56–63, 72–74.

66 This brief description of the 'labor bottleneck' borrows heavily from the lengthier analysis in Vansina, *Paths in the Rainforests*, 86. The local (and historic, it is argued) expression of this phenomenon is described in Wilks, *Forests of Gold*, 56–63, 72–74.

67 S. M. Quartey, "Development of the Human Geography of Ahantaland before 1600," *Ghana Social Science Journal* 4, 1 (May 1977), 117–27.

The northward expansion of this yam farming system was similarly constrained by immutable environmental realities. Yam gardeners found most of the lands drained by the Volta to be deficient in the nutrients required by the improved cultivars. Most of the Volta Basin had only poor laterite soils, recognizable by their pale color. These soils were not only the poorest in the area, but they were also heavy to work, became waterlogged during the region's single long wet season, and then turned quickly to cracked earth with the turn of the year's single long and intensely hot dry season.[68] Yam planters did find well-watered, rich areas in the narrow strips of alluvial soils typical of riverbanks. But they discovered that these soils, although nutrient-rich, were poorly suited to yam culture, as they were prone to annual inundation that rotted the planted tubers. Cultivating foragers continued to prefer settlement sites near potable running water and they intensely foraged the types of seedy grasses that flourished there. However, potential garden spots with rich soils and which were well watered but without flood risk were quite rare. They did find a few pockets of dark reddish-brown soils in the Volta Basin suitable for their crop culture. These widely dispersed plots were found north of the ecotone heartland of experimentation, including some parts of the Afram Plains, in the Banda Hills, a narrow band between the White Volta and the Daka River, and the lands around the Gambaga Ridge.[69]

To make a living there, cultivating foragers needed to modify their agricultural techniques to the specific local conditions. The most immediate challenge was the brevity of the rainy season on the open savannas. For garden keepers this meant a brief growing season. Moreover, the timing of the rains and their annual cumulative amounts varied greatly.[70] Yam planters sensed perpetual insecurity, as anything less than a good yam yield would fail to sustain farming communities until the next year's harvest. In choosing which yams to plant, farmers probably favored adaptability to varied rainfall over taste and texture. To the north, foragers chose to plant White Yams (*D. rotundata*) that, although of bland taste, did produce large tubers and was better adapted to short growing seasons than the sweeter, but smaller, breed of Yellow Yams (*D. cayensis*).

[68] Ahn, *West African Soils*, map 7.4 and discussion, *passim*; Boateng, *Geography of Ghana*, 59–60.

[69] Ahn, *West African Soils*, map 7.4 and discussion, *passim*; Boateng, *Geography of Ghana*, map 14b.

[70] Boateng, *Geography of Ghana*, 75; Ofori-Sarpong, "Rainfall Reliability over Ghana."

In these productive pockets, savanna yam farmers found themselves settling as neighbors of small numbers of highly mobile foragers primarily dependent on hunting, trapping, and collecting wild foods. In the years after 4000 BCE these grassland foragers gradually sought food security by keeping livestock.[71] They may have chosen to rear some tame animals, including breeds of goats (*Capra hircus*), sheep (*Ovis aries*), and cattle (*Bos indicus*).[72] Foragers must have found that they could graze cattle and sheep on dryland grasses that grew on poor soils there. These grasses provided nothing suitable for human consumption themselves but were sufficient to support small numbers of livestock—albeit in a lean and wiry state that yielded neither milk nor meat in ample quantity.

But livestock needed to be constantly monitored. As monitoring was a time-consuming but not an effort-intensive task, herding may have fallen upon adolescents and children. Cattle and sheep required some protection from wild predators when they were turned out to graze. These two creatures needed herding between areas of good feeding during the day and a return trip to the safety of the settlement in the evening. Sheep, in particular, needed to be shepherded around as otherwise they nibbled grasses down so short to the ground that they could kill their pastures. In contrast, goats required minimal attention. They were tended closer to home, in and around settlements, where they ate refuse and generally browsed without the need for people to collect fodder or protect them.

Grassland foragers were able to integrate pastoral aspects into their subsistence strategies without substantially disrupting their reliance on collecting wild foods. All three animals suited the mobile residential strategy that foragers' employed through the course of the seasons; they need not be carried from site to site. Livestock thus required only

---

71 The history of livestock in Ghana remains little studied. The following extrapolates creatively from Central and Eastern African historiography, notably: David Schoenbrun, "We Are What We Eat: Ancient Agriculture Between the Great Lakes," *Journal of African History* 34, 1 (1993), 1–31; and Vansina, *Paths in the Rainforests, passim*.

72 The history of goat and sheep are in Juliet Clutton-Brock, "The Spread of Domestic Animals in Africa," in Shaw, *et al.*, eds., *Food, Metal, and Towns*, 61–70; and Roger Blench, "Prehistory of African Ruminant Livestock, Horses and Ponies," in Shaw, *et al.*, eds., *Food, Metals, and Towns*, 71–103. The history of *Bos* in Africa is contested. The best review of the archaeological debate is Caroline Grigson, "An African Origin for African Cattle?," *African Archaeological Review* 9 (1991), 119–44. Claims of the early presence of cattle in Ghana are criticized in Stahl, "Holocene Archaeology of Ghana," 72. For the insights from genetics on the continental history of livestock, see Diane Gifford-Gonzalez and Olivier Hanotte, "Domesticating Animals in Africa: Implications of Genetic and Archaeological Findings," *Journal of World Prehistory* 24 (2011), 1–23.

moderate labor investment. They did, however, provide a dependable source of nutrient-rich foods for internal consumption, and furthermore which were a potentially profitable item for exchange with other communities. In general, cattle-keepers did not slaughter their stocks for meat. Instead they drew blood and fresh milk (the leftovers of which could be made into soured dairy foods) without endangering their herds' health. In this sense, livestock provided foragers with a cheap, renewable resource. Livestock could also be traded externally to non-pastoralists for goods and to forge social relationships outside their own communities.

No amount of human attention to their livestock could prevent losses to fatal fly-borne diseases in the humid areas of the savanna. Losses encouraged pastoralists to mitigate disease by selectively breeding long-surviving specimens towards breeds with extended life spans. This effort produced the N'dama type which better tolerated most of the endemic fly-borne diseases, including trypanosomiasis. Successive local breeding of the N'dama type produced the West African Shorthorn breed, a slight and extremely docile animal which was trypanotolerent and was well suited to human handling and survived on the limited pasturage of forest margins and clearings.[73]

Foragers also incurred hidden, indirect costs and benefits in keeping livestock. Keeping cattle and sheep probably reduced the numbers of ruminant game available to hunt. Their livestock competed with wild herds for the best grasses, and if not closely monitored, livestock could overgraze pastures and leave them bare to erosion, and thus long-term degradation. Additionally, kept animals may have also endangered their wild cousins with disease. Introduced stock probably brought new hoof and respiratory diseases that compromised wild game. There were also hidden benefits to keeping stock. Livestock tended to help the foragers' favored trees more than damaging them. While livestock might browse on the leaves of low-hanging branches of established trees, they also dropped soil-improving manure there while eating or while seeking shady refuge and chewing their cud during the heat of the day. Cattle keeping produced 'islands' of tree stands amid savanna pastureland.

After 2,000 BCE pastoral-foragers and fisherfolk on the northern and southeastern coastal grasslands began to add new rudimentary

---

[73] Ghanaian types of West African Shorthorn stand about 200 kilograms in weight at adulthood. On local types and their breeding, see Marleen Felius, *Cattle Breeds—An Encyclopedia* (Doetinchem: Misset, 1995); and Blench, "African Ruminant Livestock."

cropping to their food strategies. They adopted some food plants that Sahelian farmers had perfected to those drier environments, weak soils, and short growing seasons over the previous millennium, including varieties of pearl millet (*Pennisetum glacum*) and sorghum (*Sorghum bicolor*). Both crops were highly developed to tolerate drought and high air temperatures during the growing season. Grain planters used the new crops to extend farming techniques to dryland environments and on alluvial soils.[74] In the grasslands, where rains fell only for a few short months, these grains were quick to mature within that wet season. These grains also tolerated heavy downpours that fell on the northern grasslands. Because their starches were above ground, they could tolerate temporarily waterlogged soils, whether from rains or flood rise, without suffering damage to their edible parts. Farmers planted a broad variety of types of millet and sorghum, which were adapted to local differences in environmental qualities.[75] Where conditions allowed, they chose to plant millet rather than sorghum. Farmers found sorghum to be difficult to digest. Those who ate much sorghum (and especially the harder, red types) suffered stomach upset and diarrhea, as well as hidden weaknesses like anemia.[76]

Farmers planted varieties of sorghum and millet to ripen at different times by selecting specific varieties to local environmental conditions and staggering their sowing. Multi-staged ripening allowed grain farmers to harvest fresh grains in smaller batches over a longer period of time without the need to recruit labor in the large numbers that a single crop would have required to harvest. But farmers needed to harvest grains when they were ripe, less they be lost to birds. Grain farmers could process and eat their harvest then or they could store whole grains. In general, harder grains stored better because they were more insect resistant.

Farmers threshed, winnowed, and then processed their grain into broken meal. Most processors pounded grain in a wooden mortar, but probably some more sedentary people also used heavy grinding stones to make fine flour. Farmers could prepare the flour using various culinary techniques to produce a wide variety of starchy dishes. Perhaps the most common manner was a 'porridge,' but cooks probably deployed much more

[74] Andah, "Identifying the Early Farming Traditions in West Africa."

[75] The preference for cropping a wide genetic diversity in grain is common throughout the drylands of West Africa; see Randi Haaland, "Fish, Pots, and Grain: Early and Mid Holocene Adaptations in the Central Sudan," *African Archaeological Review* 10 (1992), 43–64; and McIntosh, "The Pulse Model," 207.

[76] J. R. J. Rowland, *Dryland Farming in Africa* (London: Macmillan, 1993); Harlan, "The Tropical African Cereals," 53–57.

complex techniques than just soaking grains in water, including using fermentation and seasonings to make tastier and more nutritious dishes.

As with cattle, there were southerly limits to the range of these grain crops. While millet and sorghum would sprout and grow to the fruiting stage in humid environments, they suffered from panicle diseases rife in very wet air. With time, farmers in wetter niches succeeded in breeding a moisture-tolerant type of sorghum (the so-called 'guinea' variety) suitable for such environments, in which the seeds twisted nineteen degrees inside the glume to vent excess moisture. But this hybrid had hard red grains, which made grinding them and preparing them into food more difficult. Farmers found these hybrids to have a particularly distinctive, bitter taste because they were consistently high in tannins. On the other hand, this chemical composition made them extremely suitable for beer making. At some early date brewers refined fermentation processes to fashion beer.

### *Initial Farming from 500 BCE*

By 500 BCE, farmers everywhere had elaborated their garden systems to the point that their plots were central to their strategy to sustain themselves. Still, farmers continued to forage, and foraged foods continued to contribute significantly to farmers' diets. Even the most successful farmers prized wild fruits and bush meat. And farmers continued to resort to foraged foods before their crops were ready to be harvested, and as a buffer against famine when their farmed crops failed. But farmers' commitment to cropping on fertile and well-watered lands was matched by the specialization of other communities in dry and infertile niches that pursued fishing and pastoralism as their primary strategies. Environmentally distinct strategies stimulated dynamic relationships between farmers and fisherfolk and herders, especially in areas of ecological transition. Exchange between these communities enhanced long-term food security for everyone, by spreading short-term risks posed by unpredictable droughts and diseases that afflicted these specialized strategies differently. Cooperation, as much as competition, defined farmers' relationships with their neighbors.[77]

In the years after 1 CE, farmers constructed a regional web of exchange that integrated the diverse resources—things, people, and ideas—that

[77] Austen, *African Economic History*, 20–21; McIntosh, "The Pulse Model."

they had adapted to the various ecological zones. Farmers built the new, regional web atop the older, mature structure of local trading networks which connected farming communities in adjacent niches, and which tied agriculturalists to non-farming people. The new network was innovative in its expanded geographic breadth. It integrated far-flung areas into a single system. Where exchange in the older system was selective and casual, the new regional system was comprehensive and formalized.

Local farmers thus accessed new agricultural resources through this network. They obtained foreign crops to sow and new domestic creatures to rear. They accessed the intellectual and practical knowledge about these innovations, and they proceeded to begin their own experiments. As always, farmers were keenly interested in new technologies, but cautious about recklessly centering their livelihoods around unproven methods. So, farmers experimented with such novelties and devised ways to integrate them into their existing subsistence strategies. The broad spectrum of farmed foods provided agriculturalists with a stable, year-round food source that reduced seasonal movements to collect wild foods. Accordingly, farmers in every niche increased their specialization in certain foods and techniques. They became more sedentary and they founded villages of greater population size and permanence. With increased sedentarism, farmers chose to accumulate more material things, including new agricultural tools and a wide variety of ceramic pots suited to cooking more elaborate meals.

With increased sedentarism, cultivators-turned-farmers began to manufacture and accumulate more handicrafts, including heavy, non-portable items. As early as 3500 BCE some had begun to fashion clay pots, and by 2000 BCE farmers everywhere were using earthenware.[78] Potters found clays suitable for their craft at river bottoms everywhere. As clays were widely distributed, there was little need to trade ceramics. For so long as communities retained a high degree of mobility, ceramic pots were a cumbersome alternative to bowls made from lighter materials, like calabashes or carved wood. More mobile people simply decided not to make or use clayworks.[79] Among more sedentary communities, potters modeled early ceramics to complement the functions of bowls made from other materials.

---

[78] For the geographic sequence and timing, see Stahl, "Holocene Archaeology of Ghana," 71; Anquandah, *Rediscovering Ghana's Past*, 29, 53, 60–61; Posnansky, "Archaeology and Linguistics," 257.

[79] McIntosh and McIntosh, "West African Prehistory (from 10,000 BC to AD 1,000)," 607.

Unglazed pots kept drinking water cool, by allowing some water to seep through the clay to the outer edge of the pot, where it evaporated. More significantly, ceramic pots withstood the heat of cooking fires. Cooks using ceramic pots could thus substitute heat energy for elbow grease in order to soften foods. Moreover, cooked food afforded eaters with better nutrition.[80] Cooks used ceramic pots to experiment with new culinary styles, notably to include simmered sauces and soups, and to turn otherwise unpalatable raw resources into delicious food.

Farming communities embraced regionalization very rapidly. The quick pace of this regionalization is apparent in the transformation of the items farmers unintentionally left behind for archaeologists to find. So rapid was this regionalization that often it has been often mistaken for the incursion of a technologically superior immigrant population that completely decimated (or assimilated) the indigenous people. But this integrated network was comprised fully of indigenous populations rather than intruders. The observed 'break' in the decorative styles of pottery and similarly abrupt (assumed) changes in residential patterns and other lifeways need not be explained simply by a migration of technologically superior people from the north. Though evidence of such changes does seem to ripple southwards over time, the same phenomenon is better described by indigenous farmers, foragers, and fisherfolk progressively transforming themselves, by selecting and embracing new crops, tools, and techniques into their subsistence strategies.

The catalyst for this regionalization was the smelting of iron. Unlike potting clay found along the lower courses of most rivers, good sources of quality iron ore were rare.[81] And unlike pottery making, smelters needed to accumulate substantial wealth and expertise in order to manufacture their product.[82] To make iron, smelters needed to procure quality ore, collect hardwoods and distill that into charcoal, and accumulate a team of trained hands to build a furnace, stoke the fires, monitor the smelt through its successive stages, and generally encourage the work through the many long hours it took to produce a mass of metallic iron. Where good iron ore was to be found, wealthy persons eagerly established manufacturing centers where smelting was done. Daboya, an ancient source of mineral salt on the White Volta, and

[80] Vansina, *Paths in the Rainforests*, 57.

[81] Cf. Anquandah, *Rediscovering Ghana's Past*, 67. Sources of ferrous laterite were everywhere, but this low quality ore was unsuitable for early smelting techniques.

[82] Austen, *African Economic History*, 22–23.

New Buipe became major iron making centers in the north. In the Banda Hills, smelters and smiths clustered in several settlements in the critical transition zone between forest and savanna environments,[83] a ecological opportunity also exploited on the southern transition zones along the central coast.[84]

Iron smelting was a great technological achievement. But for many centuries after iron first came to be smelted in the area, few iron tools fell into agricultural use. To farmers, iron tools offered only negligible advantages over cutting and digging tools they already knew how to make from familiar materials—stone and wood—which were always abundant and much cheaper to obtain. Iron's greatest initial impact was therefore to be found in the way people fostered exchange using the metal. Iron manufacturing sites became trading centers from which smelters and smiths traded unworked iron and iron wares with farmers (and pastoralists and foragers) for their food, fuel, and labor.[85] But farmers were slow to put iron tools into agricultural use. At least initially, farmers who obtained iron probably preferred to exploit its novelty by trading it and thus build their own personal networks, rather than shape the metal into expensive tools with limited agricultural benefits.

Farmers did not find iron tools to be necessary to make a living, neither to create gardens in the forest nor to reap crops. Nor did farmer's eventual acceptance of iron tools result in major changes in their subsistence strategies.[86] Perhaps the only farmers who saw early benefits in iron tools were those who sowed grain on the heavy alluvial soils in the Volta Basin.[87] In the forested areas, until the era of chainsaws,

---

[83] The area's sites are described in Stahl, "Holocene Archaeology of Ghana," 79–84; Anquandah, *Rediscovering Ghana's Past*, 67–71; and P. L. Shinnie and F. J. Keese, *Archaeology of Gonja, Ghana: Excavations at Daboya* (Calgary: University of Calgary Press, 1989). While there are many sites with ceramics of inferred 'iron age' characteristics, there are too few archaeological sites with evidence of smelting and smithing to describe the geographic outlines of this metallurgical history with precision; Duncan E. Miller and Nikolaas J. van der Merwe, "Early Metal Working in Sub-Saharan Africa: A Review of Recent Research," *Journal of African History* 35, 1 (1994), 1–35. See also Stanley B. Alpern, "Did They or Didn't They Invent It?: Iron in Sub-Saharan Africa," *History in Africa* 32 (2005), 41–94.

[84] Gérard L. Chouin and Christopher R. DeCorse, "Prelude to the Atlantic Trade: New Perspectives on Southern Ghana's Pre-Atlantic History (800–1500)," *Journal of African History* 51, 1 (2010), 123–45.

[85] Anquandah, *Rediscovering Ghana's Past*, 71.

[86] Roderick McIntosh and Susan McIntosh, "Recent Archaeological Research and Dates from West Africa," *Journal of African History* 28, 1 (1987), 1–25; Vansina, *Paths*, 58–61. Cf. Posnansky, "Early Agricultural Societies in Ghana," 150–51.

[87] Stahl, "Holocene Archaeology in Ghana," 79.

farmers preferred to ring trees (i.e. cutting a continuous break in the phloem around the trunk, to deprive the tree of nutrients those inner bark tissues transported) in order to kill them, rather than using an iron axe to fell them. And axe heads made of sharpened stone had a cutting edge that was just as effective as shaped iron. Moreover, yam farmers who experimented with iron tools also saw a distinct disadvantage in using them to harvest their crops. Iron hoe blades were harder than the wooden digging sticks that harvesters had previously used, and so were more likely to scar the root bulb's delicate protective surface, thereby exposing the starch inside to spoilage.[88] Forest farmers probably began to use iron tools extensively only when iron came to be plentiful and cheap.

From 1,000 CE, farmers everywhere began to intensify their agricultural endeavors. Although populations were small,[89] they were beginning to grow beyond the ability of existing technologies and methods to sustain them.[90] In the face of food scarcity, farmers created more hierarchical forms of labor organizations to invest more effort into cropping, and they introduced new food crops and animals to their farms.

Early in this era farmers began to breed chickens (*Gallus gallus*), a creature originally bred in East Asia. Some time after 1 CE, chickens were being reared in the Holy Lands of the Middle East, and they turn up in West Africa by 500 CE.[91] Chicken rearing followed the courses of long-distance trade that stretched across ecological zones, from the *Sahel* to West African coast. Everywhere chickens went, they introduced themselves to farmers vocally and were thus known in local language as forms of the onomatopoetic *koko*.[92] Farmers let their hens roam freely, to scratch out a living for themselves on village grounds. Hens raised by grain farmers gleaned sustenance from the scattered refuse of winnowing, but had to be deliberately provided with nutritious feed if they were to lay eggs for intensive breeding or farmers' consumption.

---

[88] This practicality might explain the modern 'prohibition' against using iron tools to harvest yams in modern times; the prohibition is mentioned in Coursey, "Yams in Africa," 397.

[89] Posnansky, "Archaeological and Linguistic Reconstruction in Ghana," 264.

[90] Anquandah, *Rediscovering Ghana's Past*, 88.

[91] The timing for the Middle East is based on the observation that there is no mention of chickens in the Old Testament. But this seems also supported by the archaeology and the genetics in Africa; see Clutton-Brock, "The Spread of Domestic Animals in Africa"; and Gifford-Gonzalez and Hanotte, "Domesticating Animals in Africa."

[92] John Berry, *English, Twi, Asante, and Fante Dictionary* (London: [privately published], 1960), 23.

In this era, farmers who worked wet soils in river valleys and savanna flood plains began to plant crops of African rice (*Oryza glaberrima*).[93] To local farmers this was a new crop, but one which bore semblance to the familiar seedy grasses which flourished along river courses. Farmers on the floodplains of the inland Niger Delta had adapted the seedy grasses growing there into a staple crop by 1500 BCE; farmers in the Senegambia area did the same by 800 BCE. Rice growing made early inroads along the northern river floodplains, where farmers found that rice yielded greater volumes of grain than millet or sorghum planted there. In the heavily forested Tano and Pra River basins, farmers planted rice crops on swampy lands that were unsuitable for any other crop, but on which rice flourished. On these plots, farmers sowed rice in advance of the annual floods. Rice farmers successfully developed techniques to maximize their rice harvest while minimizing the labor needed throughout the growing season. Their most important tool was regulating the water flow by restricting the outflow of water from these spots. Rice seeds germinated with the arrival of the rising waters, and rice plants flourished in flooded soils while submerged weeds suffocated.[94]

In the forest, farmers pioneered new areas of the forest where they intensified their cropping and collecting strategies. In this era, so-called 'big men' (Akan, *abirɛmpɔn*) coerced and organized dependent labor to do the arduous work of clearing new gardens of brush and then rearing food crops.[95] 'Big men' also laid claim to their dependants' labors outside of the agricultural season, to have them collect valuable forest products. Farmers collected forest gold, they gathered the stimulating nuts of the kola tree (*Cola nitida*), and tapped several species of palm tree for their sweet, fermentable sap.[96] Farm laborers scoured collected alluvial washes

---

93 Purseglove, "Origins and Migration of Crops in Tropical Africa," 294, 302; Harlan, "The Tropical African Cereals"; Claude-Hélène Perrot, "Semailles et moissons dans la région d'Assinie vers 1700," *Journal des Africanistes* 59 (1990), 18, 18n7. For the precolonial history of rice in Africa, see (among others) Yasmine Marzouk, ed., *Génie rural et génie paysan: sociétés rurales et techniques hydraliques en Afrique* (Paris: Éditions de l'École des Hautes Études en Sciences Sociales, 1989).

94 Water regulation in swamps and floodplains is described in Fairhead and Leach, *Misreading the African Landscape*, 104, 123–24, 140. Application of similar techniques locally is recalled in the modern saying in Akan, 'Rice likes to have wet feet.'

95 McCaskie, *State and Society in Pre-Colonial Asante*, 42–58, 275; Wilks, *Forest of Gold*, 41–90.

96 Several species were tapped for the sweet juice that flowed from their male inflorescence, including the preferred Wine Palm (*Raphia* spp.) and the oil palm. For a complete list, see Abbiw, *Useful Plants of Ghana*, 50.

for gold after the onset of the major rainy season. This prospecting season coincided with the period when gardens required little attention. Tending of kola arbors was restricted to the northern, drier parts of the forest. There, women and children collected the nuts with which their coercive bosses would strengthen commercial ties with Muslim trading communities to the north. As palm wine was always quick to spoil, it could never be traded over long distances. Therefore it remained an item to redistribute locally to build ties with the domestic community of farmers and their ancestors.[97] Even with intensification of food production, famine regularly visited farming communities in the forest. In historically specific cases of severe and sustained local food deprivation, forest farmers began to emphasize birth spacing, and perhaps practice contraception, abortion[98] and infanticide.[99]

### *Mature Farming, Circa 1400 CE*

By 1400 CE, small communities of farmers thinly populated the entire area. The starchy component of their diets (and the caloric and cultural significance of it) varied greatly, mirroring the range of local environmental niches. Their scattered village settlements probably attracted new persons and perhaps spurred imitation. But equally, their sedentary farming was balanced by other communities' increasing specialization in hunting, fishing, collecting, and trapping in areas unsuitable for the farmer's preferred crops and animal keep. Farmers everywhere pursued a common agricultural strategy, or 'tradition,' that was basic throughout the study area. They sought to minimize the amount of labor needed to produce food, seen in their aversion to gross modifications of the natural environment to make gardens and in their methods of preparing food to be eaten. The farmers also minimized and spread risk by fostering genetic diversity, by using a wide range of food crops and by farming microenvironments (like hillsides) that offered a wider range of soil types and moisture to counter annual variability of moisture and sunlight needed to

[97] Emmanuel Akyeampong, *Drink, Power, and Cultural Change: A Social History of Alcohol, c. 1800 to Recent Times* (Portsmouth: Heinemann, 1996), 1–46.

[98] Wolf Bleek, "Did the Akan Resort to Abortion in Pre-Colonial Ghana?: Some Conjectures," *Africa* (International African Institute) 60, 1 (1990), 121–31. The ethnopharmacological resources to prevent and abort pregnancies are listed in Abbiw, *Useful Plants of Ghana*, 125–26, 192.

[99] Grottanelli, "On the Origin of Things," 88–92.

sustain their crops. They continued to rely considerably on wild and semi-wild foods that they foraged, and used these foods as a buffer against famine. They integrated the relationship between producing vegetable foods and getting meat, by trapping animal pests at garden's edge and through feeding livestock the waste of human foods. Thus they used livestock as a type of savings. Men everywhere sought to dominate the production of prestige foods, which in this era were West African yams and a type of palm wine. They organized the agricultural calendar around continual harvesting to confront chronic problems storing and transporting harvests of the bulky starches they farmed.

## *Conclusion*

From earliest times to circa 1400 CE people changed incrementally how they obtained food and what it meant to them. The slow pace of this transformation was punctuated by episodes of rapid change. First, savanna ecotone foragers intensified their strategies to collect wild foodstuffs as they moved south during a period of regional desiccation, and they specialized their older techniques to make a living in the new environments they inhabited. They integrated untested novelties into a conservative preservation of what they knew would work. After 10,000 BCE, foragers along an axis of presumably increasing population density in the ecotones specialized finding wild foods indigenous to the niches they inhabited, and this created greater labor demands—and demands for more people—to do the work needed to sustain populations, and created new opportunities to localize techniques and to adapt them to local conditions to exploit wild resources. Hence there was no single, general, un-adapted center to the foraging background to later food production.

Foragers intensified gathering to the point of incipient agriculture but without placing their very survival on the bounty of the plants they managed or identifying themselves with the starchy plants they propagated. Their increasing efforts to attend to plots of starchy plants they managed hamstrung the mobility needed to forage through the seasons for wild food. As they increased their specialization in managing food resources specific to local niches they inhabited, they hitched their livelihood on foods that were too cumbersome to transport, thus exposing themselves to seasonal shortages of starches. Finally, foragers completed an unintended and unforeseen transition to relying on, rather than supplementing wild foods with, the cultigens they managed.

CHAPTER THREE

# SEEDS OF CHANGE: EARLY AFRICAN EXPERIMENTATION WITH FOREIGN STARCHES

Beginning in the late fifteenth century, African farmers near the early Portuguese strongholds on the Gold Coast began experimenting with new crops they chose from the array of plants collected through Portuguese worldwide maritime exploration of the Americas, Asia, and other parts of the African littoral. For the Portuguese it was the law of unintended consequences at work: they sought only to provision themselves, but local farmers appropriated the seeds they sowed to pursue other changes with consequences, intended and unexpected, of their own.

These African farmers—arguably women—were characteristically *selective* and tested new crops in their kitchen gardens extensively before adopting those new crops they found to be appropriate to their needs and tastes as staples. This gradual and cautious strategy runs counter to the general understanding of an immediate 'revolution' based on the high-yielding (and therefore superior, it is assumed) American starches, maize and cassava. Neither the introduction of foreign carbohydrate crops by outsiders nor Gold Coast farmers' deliberated adaptation of these crops was an historical inevitability. And Gold Coast farmers had many other considerations in mind than the sheer nutritional efficiency emphasized exclusively by the standards of modern agronomy.

This chapter focuses on local farmers' early interest in planting only certain foreign starches—particularly maize from the Americas, plantains (originally a south-east Asian crop) they obtained from Central Africa, and Asian varieties of rice—and their early uses of these crops to analyze the historical development of new food strategies, ethnic identities, cultural calendars, and gender distinctions. This chapter argues against the long-dominant historical assumption that plantains were grown and consumed on the Gold Coast prior to Portuguese maritime introduction. Linguistic, botanical, documentary evidence all link the introduction of plantain culture to the importation of African slaves from Central Africa, where the starchy fruits were a staple crop.

*The Agro-Historical Milieu*

The first African farmers to experiment with the new crops and techniques in this era were those who lived in direct proximity to the trade factories on the coast that the Portuguese established in the late fifteenth and early sixteenth century. Portuguese caravels had begun their reconnaissance of the Africa's Atlantic coastline in the mid-fifteenth century and steadily advanced the geographic range of their contacts southward. In 1471,[1] local people met Portuguese explorers who cruised the coast that the latter named *a mina de ouro* (Portuguese, 'the gold mine') or *a costa da mina* ('the coast of the [gold] mine,' from which comes the Anglophone corruption, 'the Gold Coast') for the abundance of the precious metal the seaborne adventurers bought there. Local inhabitants probably had only infrequent encounters with European mariners through the decade following initial contact, selling gold and perhaps also providing foodstuffs.[2] The demand for the latter would have been made necessary by the long voyage from Iberia, which depleted ships' stores of edibles—even if they had been able to procure full supplies of provisions before departing. The inability to provision Guinea-bound ships in Europe amid eras of drought and food scarcity is known to have hampered at least one Portuguese departure for the Gold Coast.[3] Beginning in 1504, grain farmers in Iberia suffered a series of disastrous harvests as a consequence of royal agrarian policies. Ferdinand and Isabella's attempt to boost wool production allowed shepherds to pasture their flocks in farmers' fields, and famines soon followed throughout the peninsula.[4] From 1509, Lisbon-based suppliers of provisions for ships bound for West Africa increasingly turned to Sicilian sources for their wheat.[5]

---

[1] There remains some doubt about the precise date of initial contact; see P. E. H. Hair, *The Founding of the Castelo de São Jorge da Mina: an Analysis of the Sources* (Madison: African Studies Program, University of Wisconsin-Madison, 1994), 47–48n17.

[2] Although the sale of foodstuffs to European visitors in 1470s seems likely, it remains unevidenced. Cf. the unqualified assertions in John Vogt, *Portuguese Rule on the Gold Coast, 1449–1682* (Athens: University of Georgia Press, 1979), 7; and repeated by Harvey Feinberg, *Africans and Europeans in West Africa: Elminians and Dutchmen on the Gold Coast during the Eighteenth Century* (Philadelphia: Transactions of the American Philosophical Society, 1989), 27.

[3] Hair, *São Jorge da Mina*, 118–19.

[4] Jamie Vicens Vives, *Economic History of Spain*, trans. by Frances M. López-Morillas (Princeton: Princeton University Press, 1969), 304.

[5] Ivana Elbl, "The Portuguese Trade with West Africa, 1440–1521" (unpublished PhD diss., University of Toronto, 1986), 281, 304n42.

Ships bound for the Gold Coast were at sea for some two to three months before arriving at the their destination. Sea captains charted their courses close to shore to keep sight of land to determine their position in this era before good maps and the ability to determine longitude, and so that they might put in, if necessary, to procure refreshments or find safe harbors from storms. Once European ships reached the Gold Coast, local food was not only a necessity to sustain the traders for several months while they tried to buy gold, but was also needed to provision their return voyages. Coastal residents greeted some 70 Portuguese and Castilian ships in the decade following initial contact,[6] perhaps stimulating local farmers to expand food cropping slightly to generate surpluses to sell opportunistically. But these sporadic visitors need not have supported modifications of local farmers' basic, self-supporting strategies to provision them.

Locals intensified their relationship with these foreigners after 1482 when the Portuguese established a permanent trade-post (named *São Jorge da Mina*, 'St. George of the Mine') on the central coast at the large African settlement later known as Elmina.[7] The Portuguese chose to build at this site at least in part because they believed the town would be a source of adequate provisions and potable water.[8] Negotiations between Portuguese and local élites concerning the establishment of the fort probably included proffers of local food and official arrangements to supply the trade fort with provisions.[9] Such arrangements would have been made more urgent by the considerable size of the Portuguese work force. On the construction site some 600 people toiled under Portuguese control. Counted among them were African slaves that the Portuguese brought

---

6 On the contacts between reconnaissance in 1471 and establishment of the fort in 1482, see Hair, *São Jorge da Mina*, 48–9n18; Elbl, "Portuguese Trade with West Africa," 212, 455.

7 The name 'Elmina' is of later provenance, dating from the mid-seventeenth century; see Hair, *São Jorge da Mina*, 44n4. The toponym is used here, if a bit anachronistically, for convenience.

8 Rui da Pina, "Crónica de el-Rei D. João II" [1792] cap. 2 ('[...] *grande povoraçam, que dava espança d'agoa doce, e d'outras provisões, aas gentes pelos tempos comprimadoiras*'). English translations are in J. W. Blake, trans. and ed., *Europeans in West Africa, 1450–1560* (London: The Hakluyt Society, 1942), 70–78; and slightly differently in Hair, *São Jorge da Mina*, 16 and 71n97, which also includes a transcription of the original Portuguese and additional annotation, 99–103. The importance of provisioning in the Portuguese selection of Elmina was also noted in the earliest detailed Dutch account of the Portuguese fort; see Albert van Dantzig and Adam Jones, trans. and eds., *Pieter de Marees, Description and Historical Account of the Gold Kingdom of Guinea, 1602* (Oxford: Oxford University Press and the British Academy, 1987), 218.

9 Hair's speculation; *São Jorge da Mina*, 33–34, 37.

from the Niger Delta to assist in the construction.[10] After completion of the fort, the total number of personnel was reduced to a modest 60, but these permanent residents were completely dependent on food from outside its walls, be it imported from abroad or purchased locally.

In the first decade of Portuguese settlement on the Gold Coast, fort personnel frequently suffered interrupted supplies of food from home, despite official pledges of regular resupplies of provisions. Wheat flour, olive oil, and grape wine were the core staples from Lisbon. Sometimes these interruptions were due to poor harvests in Iberia, but more often and even systemic was the persistent lack of available ships in home ports suitable to make a resupply run to remote outposts in 'Guinea' and the poor communication system and inefficient feudal organization which plagued the early overseas empire.[11] In such situations of intermittent scarcity and hunger, fort personnel must have turned to local food supplies, thus extending the maritime strategy of procuring fresh food whenever it was available. But this alien group must have approached local foodstuffs with some trepidation. Although they were familiar with some local produce (such as sorghum, millet and rice, all of which were grown in Iberia in the fifteenth century), most other local foods were absolutely foreign to them and were therefore somewhat suspect in an European expression of the universal human propensity to identify themselves with what they ate. They thought some local foods to be incompatible with their constitution and therefore considered them to be injurious to their health.[12]

Fort personnel presumably first sought out local ingredients that bore at least some resemblance to something with which they were familiar,

[10] Hair, *São Jorge da Mina*, 58n44.

[11] Elbl, "Portuguese Trade with West Africa," 290–94; Vogt, *Portuguese Rule on the Gold Coast*, 153.

[12] Hair, *São Jorge da Mina*, 12, 71n96, citing the record of João de Barros [1552]. Hair speculated that perhaps Barros had himself suffered himself from a bout of ill health while on the Gold Coast, and had attributed this to the local foods he ate there. Europeans in the early modern period had notions about diet, nutrition, and their effect on health (based on ingredients' supposedly inherent degrees of 'heat,' 'coldness,' 'wetness,' and 'dryness') which seem peculiar nowadays; see Rachel Laudan, "Birth of the Modern Diet," *Scientific American* 283, 2 (2000), 76–81; and Nancy G. Siraisi, *Medieval and Early Renaissance Medicine: An Introduction to Knowledge and Practice* (Chicago: University of Chicago Press, 1990). By the eighteenth century, European ideas about diet had changed greatly, but expatriates continued to attribute their bad health in the tropics to consumption of local foods; see Dominique Juhé-Beaulaton, "Les jardins des forts Européens de Ouidah: premiers jardins d'essai (XVIIIème siècle)," *Cahiers du Centre de Recherches Africaines* 8 (1994), 90–91; and Karen Ordahl Kupperman, "Fear of Hot Climates in the Anglo-American Colonial Experience." *William and Mary Quarterly*, Third Series, 41, 2 (April 1984), 213–240.

and which they could process into forms they knew. They could, for example, use locally produced millet and sorghum flour to bake into bread, and they recognized that palm oil was a viable substitute for imported olive oil. It is notable that alcohol—in the form of local palm wine, although it was not nourishing—was among the first of local foods to be taken by these foreigners, probably in the vain hope of alleviating the loneliness and depression they must have felt in such an isolated and remote posting. Records show that at least by 1495–99, fort personnel were trading for local food of uncertain identification in order to supplement their diet.[13]

For European foreigners on the Gold Coast, these local foods remained always but inferior substitutes for products from home. Even if proved through collective experience that they did not harm Europeans' health, consumption of indigenous foods indicated an obvious social denigration. Bread made from local millet and sorghum flour alone could never be baked into loaves resembling European bread, which was their age-old, celebrated 'staff of life.' Flour ground from these African grains lacked gluten (long protein structures) which, after kneading arranged the gluten into layers, could trap and hold air produced by yeasty fermentation and thus expand when heated to produce the 'risen' bread that Europeans (since Pliny's classical writings) had considered to be the most healthful. Only with the addition of imported wheat flour, if any was indeed available, could the castle bakers produce something akin to European bread. But these loaves were of the less refined, hence cheaper, darker, denser sort which culinary color prejudiced Europeans associated with peasant or prisoner fare.[14]

Thus acute stress on supplies of imported provisions made food a marker of social differentiation within the fort community. High-ranking Portuguese administrators and traders at the fort could presumably insulate themselves, if only temporarily, from such culinary insult and presumed health detriment by reserving scarce quantities of esteemed, imported provisions for themselves. They could also use high-value, imported edible

[13] The early records for trade in food are analyzed in J. Bato'ora Ballong-Wen-Mewuda, *São Jorge da Mina, 1482–1637: La vie d'un comptoir portugais en Afrique occidentale*, 2 vols. (Lisbon and Paris: Écoles des Hautes Études en Sciences Sociales, collection du Centre d'Études Portugaises: Fondation Calouste Gulbenkian/Commision Nationale pour les Commérations des Décourvertes Portugaises, 1993), *passim*; see also Vogt, *Portuguese Rule on the Gold Coast*, 50, 71.

[14] Harold McGee, *On Food and Cooking: the Science and Lore of the Kitchen* (New York: Charles Scribner's Sons, 1984), 273–313. The baking of such 'dark, coarse bread' at Elmina is in Vogt, *Portuguese Rule on the Gold Coast*, 47, 49–50.

novelties—initially European livestock and alcohol—to forge social relationships with local African merchants and dignitaries who controlled the sources of gold that they sought to purchase. In 1518, the Portuguese gave goats as gifts to merchants based in Wassa, up the Ankobra River from Axim.[15] Perhaps the goats were not imported breeds, but instead local ones, and thus an example of the foreign traders working within the local value system. Privileged persons who controlled the rationing of imported provisions were less concerned with the diet of their countrymen who filled the ranks of the support personnel and soldiers. Most had been banished to western Africa from Portugal as criminal outcasts (Port., *banidos/as* or *degradados/as*); the majority were men, but a small number of women were there as well, working as cooks and, perhaps, as prostitutes.[16] It is difficult to know how to describe these Portuguese women, and later Luso-African women who filled similar positions, because the early sources tell us little about them. They were described as outcasts, slaves (both Europeans and Africans), concubines, wives, and condemned as 'whores.'[17] If these banished persons were lucky enough to survive their forced service, which few did, they tried to make new lives for themselves as petty traders as well as by marrying into local families, and thus they created a dynamic economic and cultural interface.[18] But when the fort

[15] Vogt, *Portuguese Rule on the Gold Coast*, 83.

[16] Hair, *São Jorge da Mina*, 36, 91n195–6.

[17] On the problem, see Emmanuel Akyeampong, *Drink, Power, and Cultural Change: A Social History of Alcohol in Ghana, c. 1800 to Recent Times* (Portsmouth NH: Heinemann, 1996), xvii–xix; and Adam Jones, "Prostitution, Polyandrie oder Vergewaltigung? Zur Mehrdeutigkeit Europäischer Quellen über der Küste Westafrikas zwischen 1660 und 1860," in Adam Jones, ed., *Außereuropäische Frauengeschichte: Probleme der Forschung* (Pfaffenweiler: Centaurus Verlag, 1990), 123–58.

[18] The Luso-African character of early Gold Coast imperial society is woefully undeveloped, a deficit made more apparent by the revealing work done on these communities in Senegambia and Angola; see Jean Boulègue, *Les Luso-Africains de Sénégambie* (Lisbon: Ministério da Educação, 1989); Joseph Miller, *Way of Death: Merchant Capitalism and the Angolan Slave Trade, 1730–1830* (Madison: University of Wisconsin Press, 1988). By 1600, De Marees described the Luso-African population as of both substantial size and importance to the operation of the fort; Van Dantzig and Jones, trans. and eds., *Pieter de Marees*, 36–7, 217, 220. Perhaps these Luso-Africans assimilated in later eras into the Dutch- and Anglo-African mulatto communities, whose history is now coming to be written. See Natalie Everts, "Cherchez la femme: Gender-Related Issues in Eighteenth-Century Elmina," *Itinerario* 20, 1 (1996), 45–57; Natalie Everts, "'Brought up well according to European Standards:' Helena van der Burgh and Wilhelmina van Naarssen: Two Christian women from Elmina," in Ineke van Kessel, ed., *Merchants, Missionaries & Migrants: 300 years of Dutch-Ghanaian Relations* Amsterdam: KIT Publishers / Accra: Sub-Saharan Publishers 2001), 101–109; and Michel Doortmont, "An Overview of Dutch Relations with the Gold Coast in Light of David van Nyendael's Mission to Asante in 1701–02," in Van Kessel, ed., *Merchants, Missionaries & Migrants*, 19–31.

administrators rationed scarce provisions, the social and physical welfare of these forced servants must have been of minimal concern.

Of greater, if ultimately only self-interested, concern to the traders and administrators was sustenance of the African slaves lodged at the castle. From the 1470s Portuguese traders purchased slaves from the 'Slave Rivers' of the coastal Niger Delta. In early sixteenth century they expanded their slaving operations to carry slaves from the kingdom of Benin located to the west of the Niger and, increasingly after 1513, to ship slaves from the Central African kingdom of Kongo. The annual volume of slaves sent to the Gold Coast probably ranged from 200 to 500 per year, but accurate totals have been very difficult to determine, given poor official records of legal trade and the presumably substantial size of illegal, private sales.[19]

Imported slaves needed to be fed until they could be sold locally for the gold the traders primarily sought or, less favorably, for high-grade elephant ivory. Portuguese traders recognized that the slaves they owned were an imminently perishable inventory. This bulk trade in human beings landed slaves in terrible nutritional shape after having been seized somewhere else, bulked in very uncertain circumstances, and carried on vessels not stocked or designed to sustain them. Once at Elmina, slaves left unfed or badly nourished were prone to disease and death, which made for significant losses on fort accounts and to the traders' personal balances. When the promise of sufficient provisions from home proved to be hollow and the supply of foodstuffs from the local African community became more unpredictable, the traders sought to experiment.

Their early experiments focused on reorganizing local and regional food resources. Traders and administrators attempted to supplement scarce food shipments from Europe with starchy foods from the same sources as their slave inventories. They imported yams and palm oil from the port of Bonny, the primary trade settlement in the eastern Niger Delta, and from Portuguese settlements on the islands of São Tomé and Príncipe.[20] Around 1500 the Portuguese experimented with organizational change of

[19] The most sophisticated analysis is found in Ivana Elbl, "The Volume of the Early Atlantic Slave Trade, 1450–1521," *Journal of African History* 38, 1 (1997), 31–75; and her dissertation, "Portuguese Trade with West Africa," especially pp. 339, 367–68, 471–72. See further, John Vogt, "The Early São Tomé-Príncipe Slave Trade with Mina, 1500–1540," *International Journal of African Historical Studies* 6, 3 (1973), 453–67; John Vogt, "Private Trade and Slave Sales at São Jorge da Mina: a Fifteenth-Century Document," *Transactions of the Historical Society of Ghana* 15, 1 (1974), 103–10.

[20] Elbl, "Portuguese Trade with West Africa," 456, 549; Hair, *São Jorge da Mina*, 72n98; Ballong-Wen-Mewuda, *São Jorge da Mina*, unpaginated chart ("Mouvement des naivres de traite des marchandises africaines"), facing p. 358.

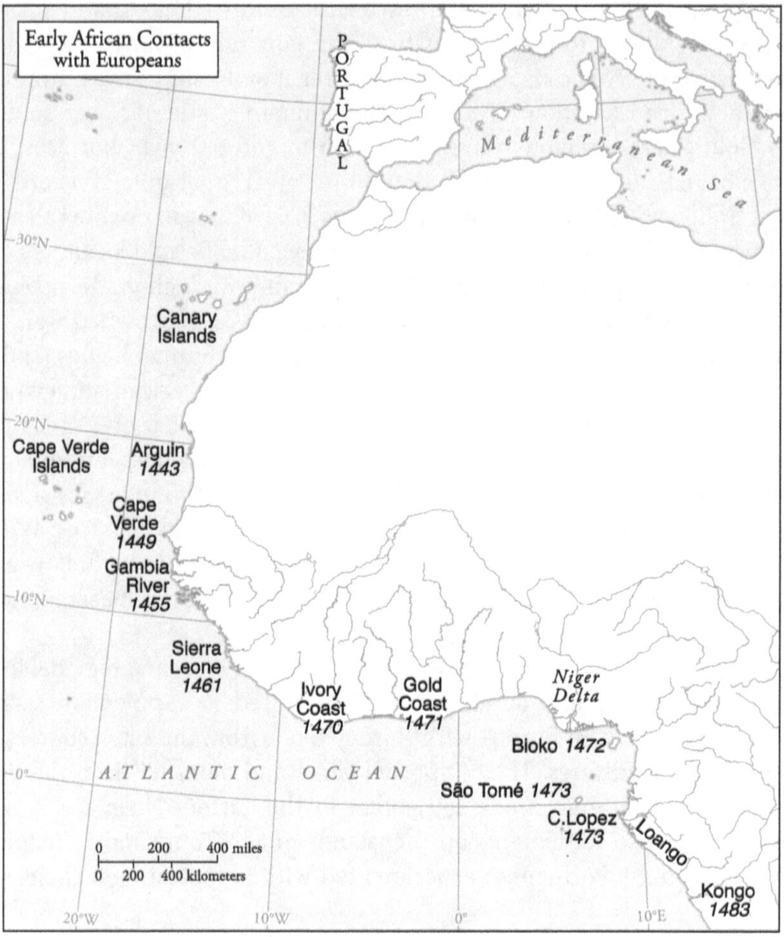


Map 5. Early Contacts in Western Africa in the Atlantic Era

the Mina factory to curtail private slave trading by fort personnel. A key aspect of this reform was to strip personnel of their periodic maintenance stipends, which some apparently used as working capital for their own slave-trading interests. But the reorganization also highlighted the need to better manage scarce food resources. The new regime directly supplied personnel with food, and a new administrative position (Port., *almoxarife dos mantimentos*; roughly 'quartermaster') was established to oversee the acquisition, storage, and dispensing of provisions. These rations included daily allotments of coarse bread baked at the fort and some imported wine, and monthly amounts of palm oil, honey, and vinegar.[21]

The provisioning needs of the Portuguese increased sharply in the first decades of the sixteenth century due to an expansion of the Portuguese presence on the Gold Coast with new trade forts established at Axim (1503, and enlarged in 1515), Shama (1523, but a site of trade since the 1470s) and Accra (date unknown).[22] Persistent and increasingly grave provisioning problems in the sixteenth century encouraged the outposts to experiment with raising some food crops and animals themselves. In undertaking food production for themselves, they sought to develop local systems that they hoped would be both cheaper and more certain than unpredictable metropolitan resupply networks. Producing their own food locally could dramatically reduce the cost of provisioning, by eliminating high transportation charges from Europe that typically doubled the cost of such bulky, but low-value goods.

Portuguese experimenters in search of cheap starchy staples initially experimented with the European grains with which they were familiar.

[21] Vogt, "Private Trade and Slave Sales," 107. It is difficult to imagine how this vinegar was to be used at Elmina, as it was not a core ingredient in Portuguese culinary traditions. Perhaps this vinegar was spoiled grape wine that could not be given as payment or tribute, so was fobbed off as a provision, or perhaps it was used to tenderize tough meat (as vinegar's acids denature proteins, albeit at the expense of drying out the flesh). Most speculatively, it might have been used or to flavor local 'greens' along with locally obtainable salt, in the contemporary European manner. Local African foodways had (and have) little use for vinegar, although it was readily available from palm wine, which spoils rapidly. Note that Akan word for vinegar (*asin*) is a corruption of the Dutch name (*azijn*). The borrowing might indicate a new technology, as people are generally loath to borrow a word for something they already have. Finally, in contrast to Africans' year-round harvesting, the relatively short, Northern European growing season stimulated development of pickling techniques to preserve vegetables.

[22] Elbl, "Portuguese Trade with West Africa," 284, 455; Albert van Dantzig, *Forts and Castles of Ghana* (Accra: Sedco Publishing Ltd., 1980), i–v, 1–9.

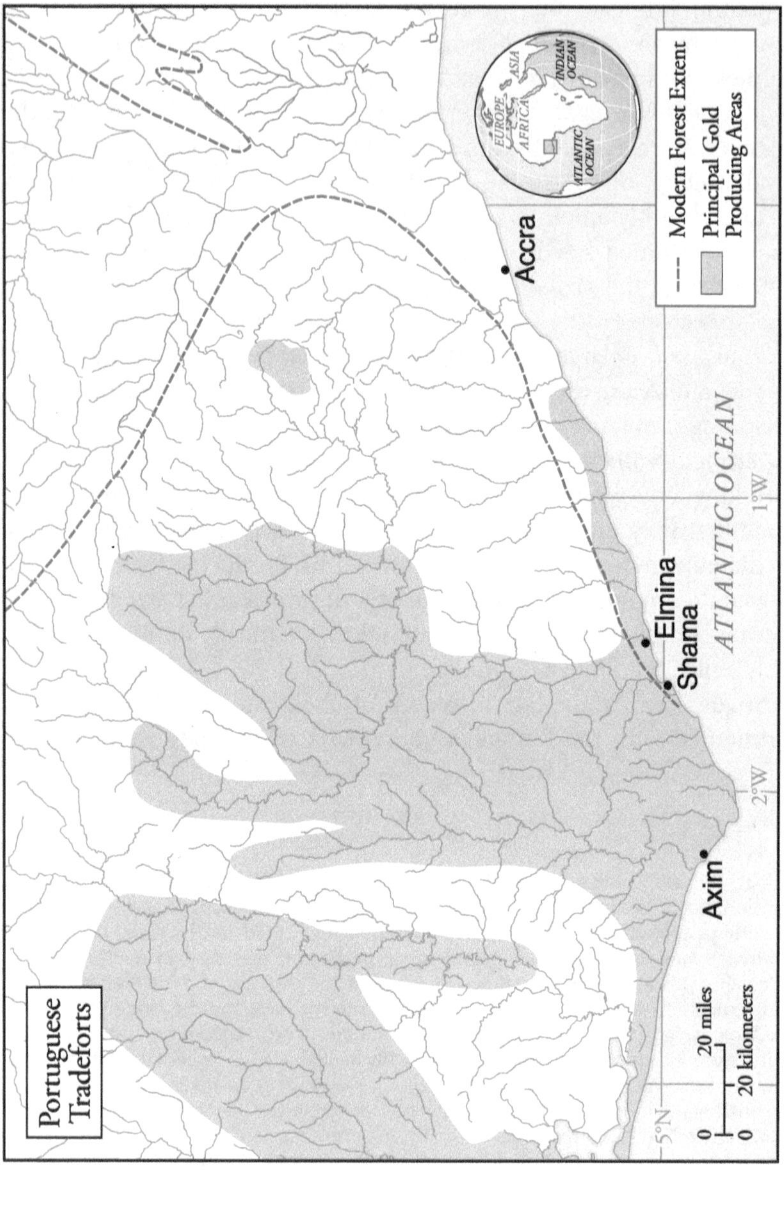


Map 6. Portuguese Tradeforts

They planted trial crops of wheat, their preferred grain, on the Gold Coast,[23] as they did in Senegambia,[24] São Tomé,[25] and Angola,[26] but this crop failed to produce grain anywhere in Western Africa. On the Gold Coast, they then may have pursued planting starchy staples without prestige in Iberia but which were better adapted to tropical climates, including Portuguese varieties of millet.[27]

If the élite expatriates were eager to sustain familiar foodways and only reluctantly force the least powerful of their lot—exiles and slaves—to eat disparaged foreign foods, wealthy Africans in the area were of a completely different mind: they were eager to experiment with foreign foods. Initially African notables acquired experimental crops and animals from the forts under the guise of 'gifts' from traders. Later, local notables paid dearly to obtain these imported natural novelties, for the Portuguese traders rightly appraised the foods to be locally unique and, therefore in local economic

[23] Hair, *East of Mina*, 81: 'All the seeds [brought from Portugal] grow, and if they do not produce crops—as a few grains of wheat failed to do, although the plants grew as tall as a man and developed very large ears…'

[24] G. R. Crone, trans. and ed., *The Voyages of Cadamosto and Other Documents* (London: The Hakluyt Society, 1937), 42. Rye and barely also failed to grow there.

[25] Alpern, "European Introduction of Crops," 14.

[26] "Regimento de D. Manuel a Simão da Silva (1512)," in António Brásio, ed., *Monumenta Missionaria africana: Africa ocidental (1600–1622)*, sér. 1 (Lisbon: Agência Geral do Ultramar, Divisão de Publicações e Biblioteca, 1952) vol. 1, 239–40. This document is unique in that it lists seeds sent from Lisbon to Africa for the purpose of experimentation. Among these seeds was wheat, and the others have been identified in Alpern, "European Introduction of Crops," 32n7. It is presumed that this wheat failed here, as it did in the other tropical locales. For further indications of metropolitan authorities' sense of urgency for better provisioning in this era, see the above-mentioned *regimento*; the 12 December 1530 letter to the *feitor* and *oficas* of São Tomé in Brasio, ed., *Monumenta*, vol. 1, 551–2; and Miller, *Way of Death*, 351.

[27] C. C. Reindorf, *History of the Gold Coast and Asante, based on Traditions, and Historical Facts*… (Basel: Basel Evangelical Missionary Society, 1895), 268. Reindorf was a pioneering local historian of mixed African and European parentage, and who drew on the different historical traditions of both of these communities (i.e., documents and oral traditions) to craft his historical reconstruction. I suspect that this particular bit of information emerged as his way of reconciling contradictory evidence: on the one hand, the (correct) indigenous impression that millet was an ancient crop of the Gold Coast; and on the other hand, his reading of Willem Bosman's account (a source he mentions as one of his literary references) in English translation, where the introduction of maize (so identified in the Dutch-language original; and the episode probably cribbed from De Marees's text) was consistently mistranslated as concerning 'millet.' To reconcile the contradiction, he posited the introduction of a 'new type'—a tactic he also used to describe the history of plantains (see below). I regret that I have not been able to check Reindorf's manuscript written in Gã, and now conserved as the Basel Missionary Archive (Switzerland), to verify that this 'millet' in the published English versions is indeed a correct translation; cf. Gã *ŋmaa*, 'millet.'

value systems, as expensive.[28] Conspicuous consumption of expensive foreign foods confirmed the privileged status of rich men,[29] and was captured in proverbial sayings like, 'a rich man eats delicious foods' (Akan, *Osikani na odi ade a ɛyɛ dɛ*).[30] Their exclusive access to alcoholic 'overseas drink' (Akan, *apo so nsa*) also enabled them to reinforce their secular power through ritual use of drink to communicate with ancestral authorities.[31]

Extant contemporaneous records throw little light on agricultural experiments in these early times. No single documentary source describes the 'moment' of introduction for any of the new crops and farm animals to the Gold Coast, nor the personal identity of the historical agents, their precise physical locations, their various motivations, or their methods. Thus many of the answers to the normal questions of historical inquiry must be inferred from documents written a hundred years hence, contemporary documents regarding similar introductions elsewhere in western Africa, comparative linguistic data, historic and modern oral traditions, and botanical and environmental conditions.

The coastal farmers who were the first to encounter foreign technologies invented names in their own languages to identify the new crops and animals. They thus created a linguistic heritage of what they had in mind evident in vocabularies in modern West African languages. In some cases, they borrowed the names they learned from those who had provided the novel agricultural technology and interpreted these 'loan words' through the phonological conventions of their own speech communities (i.e., sound shift). For example, Akan-speaking experimenters of the central coast who took to rearing European domestic swine (*Sus domesticus*), initially brought to West Africa by the Portuguese, reinterpreted the Lusophone moniker for the animal, *porco*, into their own language as *prako*.

In other cases, the experimenters interpreted the foreign items as variants on familiar categories of foodstuffs, thus revealing the structures of their thought about eating. Somewhere on the southern coast, speakers of one Akan dialect coined neologisms to name many of the new, experimental crops and domestic animals. They invented an adjective meaning

---

[28] Van Dantzig and Jones, trans. and eds., *Pieter de Marees*, 216: 'All of these [new crops] were great and strange novelties for the Negroes, who gave a lot of money for some fruits when they first came out, because of the appetite they had for them.'

[29] In modern Ghana, white bread is the penultimate 'Western food' (Nzema, *aboloba ale*); Veronica Lazzarini-Viti, "Western Foods and Traditional Diet in Ghana," *Paideuma* 24 (1978), 103–9.

[30] Christaller (compiler), *Three Thousand Six Hundred Proverbs*, trans. Kofi Ron Lange, 87.

[31] Akyeampong, *Drink, Power, and Cultural Change*, 14–15.

'from overseas'[32] (Akan, (*o*)*boro*) and joined this descriptor to names they used for familiar analogs. Akan-speakers had already begun to use the adjective-stem by 1479, when the first surviving list of local words used at Elmina was compiled,[33] and in the following decades they created neologisms of this sort to name all manner of novel items, from trade items like European glass and leather to foreign tools such as ship's sails.

These sorts of names from neologisms and loan words, when culled from the word-lists compiled by early European visitors in guidebook-like aids and from modern dictionaries provide clues about otherwise unevidenced aspects of agricultural transformation in these early times. The semantics of these neologisms describe Gold Coast farmers' initial impressions of the new crops and loan words point to economic contacts and technological exchange in the past. By comparing the forms of these names in the area's several languages it is also possible to make inferences about the sequence and geographical direction[34] of agricultural adoptions in this era (e.g., Akan *prako* > Gã *kploko*, via a standard /r/ > /l/ sound change).

The early linguistic data present special analytical challenges because foreign visitors, even those with keen ears and a real interest in mastering

32 The meaning the adjective *oboro* has often been glossed as 'European,' but that is of course an anachronistic translation here, as the idea of 'Europe' would not have been a part of the mental (or cultural, racial) geography of any Akan-speaker in the sixteenth century. As late as the nineteenth century, the local African conception of 'Europe' was yet vague; McCaskie, "Innovational Eclecticism," 42–44. The etymology of the adjective *oboro* has not yet been addressed in print, to my knowledge, beyond a short note by J. B. Danquah, the scholar turned nationalist politician and statesman. Danquah gave 'horizon' or 'overseas' as the original meaning of the Akan word *boro*, and he attributed his source to then-deceased *Akuapenhene* [King of Akuapem], Nana Frederick William Kwasi Akuffo. See J. B. Danquah, "Notes on 'Obroni' and 'Buronya,'" *Transactions of the Gold Coast and Togoland Historical Society* 2 (1952), 71–71. But it is worth noting that this meaning is not given in either edition of Christaller's *Dictionary* (1881 and 1933). I tentatively suggest that it might be related to the Akan verb *boro asu*, 'to swim' or 'move in water,' hence my preferred translation as 'from overseas.' This word has however clearly come to now mean 'European,' but there are other derivative meanings and words (e.g., any non-African foreigner, Ghanaians with light complexions, 'Western' imports or 'Westernized' Ghanaians).

33 P. E. H. Hair, "A Note on De la Fosse's 'Mina' Vocabulary of 1479–80," *Journal of West African Languages* 3, 1 (1966), 55–57; David Dalby and P. E. H. Hair, "A Further Note on the Mina Vocabulary of 1479–80," *Journal of West African Languages* 5, 2 (1968), 129–31.

34 The linguistic geography of the Gold Coast in early times is believed to have been quite similar to the modern map; see M. E. Kropp-Dakubu, "Notes on the Linguistic Situation on the Coast of Ghana during the Nineteenth Century," *Research Review*, new series, 1, 2 (1985), 192–202; and M. E. Kropp-Dakubu, "The Peopling of Southern Ghana: A Linguistic Viewpoint," in Ehret and Posnansky, eds., *The Archaeological and Linguistic Reconstruction of African History*, 245–55.

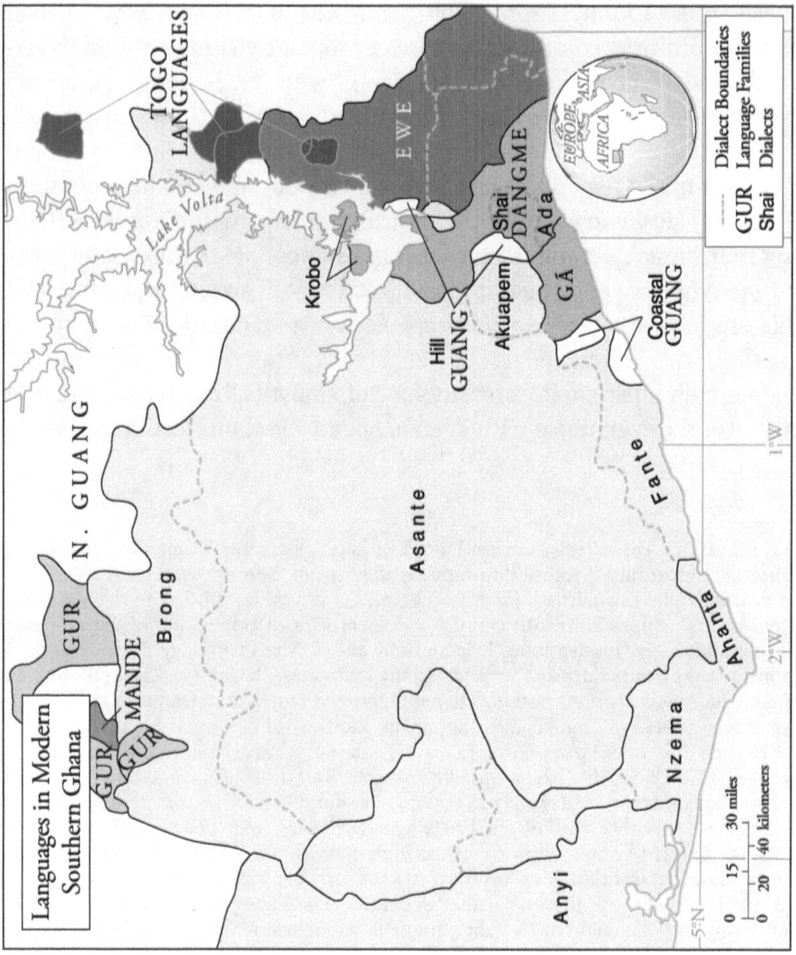


Map 7. Languages Spoken in Modern Southern Ghana

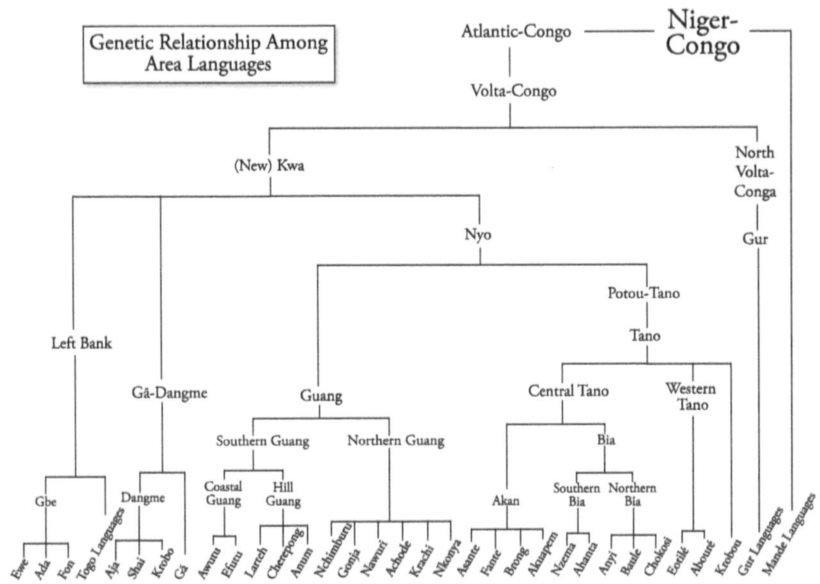

(source: adapted from Bendor-Samuel 1989)

Illustration 1. Genetic Relationship among Area Languages

the local vocabulary of trade, recorded word lists amateurishly, and in the wildly differing orthographies of their own various mother tongues and personal idiosyncrasies.[35] These data are further complicated by the early visitors' employment of local translators. From the earliest Portuguese times, traders employed (or enslaved) locals to guide them in their contacts on the linguistically heterogeneous Gold Coast,[36] and at least by 1600, some African merchants made it their practice to bring along their own interpreters when meeting with Europeans.[37] Early word lists often therefore show several degrees of removal from the original speech, as it moved from speaker to oral translator to European writer.[38] The development of a traders' coastal pidgin based on Portuguese eased inter-continental and intra-regional communication in early times (and continued to be used

[35] P. E. H. Hair, "Collections of Vocabularies of Western Africa before the Polyglotta: a Key," *Journal of African Languages* 5 (1966), 208–17.

[36] Elbl, *Portuguese Trade with West Africa*, 209; Hair, *São Jorge da Mina*, 85n152.

[37] Van Dantzig and Jones, trans. and eds., *Pieter de Marees*, 33, 46–50.

[38] As Adam Jones concluded, 'Die Westafrikaner selber kommen bestenfalls in der Form einer indirekten Rede zu Wort, in der in einer europaïschen Sprache berichtet wird, was ein Afrikaner angeblich gesagt hat'; in Adam Jones, *Zur Quellenproblematik der Geschichte Westafrikas, 1450–1900* (Stuttgart: Franz Steiner Verlag, 1990), 91.

on the Gold Coast well past Portuguese withdrawal from the area, and at least until the middle of the eighteenth century) but this now further complicates the early word lists with some seepage of later pidgin terms for imported technologies in place of the indigenous synonyms originally used.[39] In all, comparing sound patterns in these lists by themselves cannot support fine lines of historical resolution in many instances. However, a broad base of old and new names can indicate patterns of agricultural change in early times, including the history of experimentation with the starchy staples local farmers named 'overseas yam' (Akan, *oborode*) and 'overseas grain' (Akan, *oborowi*),[40] and another new starchy crop, Asian rice, which seemed so familiar that farmers did not think it needed a new name.

## *Plantains*

In the first decade of the sixteenth century, or shortly thereafter, local farmers began to experiment with cultivating and consuming plantains (*Musa* spp. AAB[41] [see Illustration 2; page 77]), a crop they named 'overseas yam' (Akan, *oborode*). Although to the eye this foreign crop was quite unlike yams—it was a perennial plant that produced above ground fruits, rather than an annual, herbaceous vine that yielded a tuber—coastal farmers recognized several attractive similarities between the new and old crop. They found that they could reproduce plantains vegetally as they

---

[39] Noted in 1662–69 by Wilhelm Müller in Jones, trans. and ed., *German Sources for West African History*, 183, 247; and as late as 1739 by Ludewig Rømer in Selena Axelrod Winsnes, trans. and ed., *A Reliable Account of the Coast of Guinea (1760) by Ludewig Ferdinand Rømer* (Oxford: Oxford University Press for the British Academy, 2000), 32–33 with note 7.

[40] I recognize that in many modern dialects of Akan, the historic forms of the neologisms have been adapted over the last five hundred years into current forms that can potentially obscure their original form, as it is argued here. Word lists compiled throughout the Atlantic era indicate aspects of these transformations; see word lists, below. Typical changes include alteration of the noun stem (e.g., *oboro* > *aboro*) and abbreviation of the second vowel sound (*aboro* > *abro*).

[41] A brief primer on *Musa* genomics is warranted here. *Musa* plants belong genetically to three families, and two of these are found in Africa: AAB and AAA. These acronyms refer to specific triploid combinations of genetic contributions from *M. acuminata* (A) and *M. balbisiana* (B), the uncultivated Asian ancestors of the edible *Musa* families. In colloquial anglophone terminology, *Musa* AAB is 'plantain,' a starchy fruit that requires cooking, and *Musa* AAA is the sweet, ready-to-eat fruit—but there are some cultivars that contradict this informal division. *Musa* AAB is further divided into groups defined by the morphology of their bunches, ranging from the most compact French types, through French Horn to False Horn, to the least compact Horn plantains.

did with yams, and they could harvest from the new crop a starchy fruit which they could prepare using the same techniques that cooks already applied to yam tubers in order to make the starchy dishes that accompanied most meals: they could boil the peeled fruits in water and then crush them with red palm oil to make *edo*, or pound them unmixed into *fufu*, or they could boil sliced fruits to make *ampesi*.

Just when these farmers began their plantain experiments is unknown, but the historical fact that plantain culture in the area of the Gold Coast post-dates initial European contact with this area is certain. This argument contrasts with the prior historiographical understanding that plantain culture in this area is, as in the rest of the continent, very ancient. However, documentary, linguistic, cultural, ecological, and agronomic information points strongly towards an early sixteenth-century beginning of plantain culture on the Gold Coast and, indeed, all of West Africa.

Why had Gold Coast farmers not begun before then to grow and eat plantains, a crop that farmers of the moist forests of East and Central Africa had put at the center of their foodways for a long time, for several hundred years if not a millennium?[42] Quite simply, Gold Coast farmers had probably never acquired the botanical means with which to experiment with plantains before the early sixteenth century. The low, rolling grasslands known as the Dahomey Gap constituted a natural barrier to the westward spread of plantain culture from the wet forests east of the River Niger. Farmers of this grassland 'gap' had presumably seen their own experiments with plantains fail due to low environmental moisture there, which was detrimental to this water-loving plant; or they had rejected it in favor of more familiar dryland crops, inclduing sorghums, millets, and some varieties of African yams. Thus when Portuguese caravels first began to survey the West African coast, no observer[43] caught sight of plantains:

---

[42] Precise dates for plantain culture are also lacking in East and Central Africa, as they are in West Africa. Gerda Rossel presented authoritative inferences from comparative lexical and botanical data about this crop's spread throughout the continent; see Gerda Rossel, *Taxonomic-Linguistic Study of Plantain in Africa* (Leiden: Research School CNWS Publications, 1998). For regional descriptions of these same historical contours but written in humanistic and historical terms see for Central Africa, Vansina, *Paths in the Rainforests*, 61–65; and for East Africa, David Lee Schoenbrun, *A Green Place, A Good Place: Agrarian Change, Gender, and Social Identity in the Great Lakes Region to the 15th Century* (Portsmouth NH: Heinemann, 1998).

[43] *Contra* J. W. Purseglove, "The Origin and Migrations of Crops in Tropical Africa," 295: 'They [*Musa* spp.] spread into the heart of the continent via the Zambezi valley and the Great Lakes and then across the Congo to West Africa, where they were found by the earliest Portuguese explorers.'

not Alvise da Ca' da Mosta,[44] who visited the Senegambia regions in two voyages in 1455 and 1456, nor Eustache de la Fosse,[45] whose trade mission to western Africa was interrupted by his capture and forced service at Elmina in 1479–80, nor any of those Portuguese mariners whose accounts of western Africa circa 1506–10 were collected and published by Duarte Pacheco Pereira.[46] Even if Gold Coast farmers had heard of plantains (or, more unlikely, had received their fruits via some sort of long-distance sea and lagoon canoe trade) before the late fifteenth century, existing transportation possibilities were not sufficient to deliver the specific genetic material with which they could successfully experiment.

Those Europeans who sought to introduce plantains into the alien environment of the Gold Coast must have found the crop a much more challenging—perhaps even frustrating—experiment than were familiar seed crops from Europe. Dried seeds were easy to obtain, and inexpensive to store and transport, they were light, small, and by the nature of a seed's function they remained in dormant state indefinitely so long as they were kept reasonably dry and whole. Of course, an experimenter's success with seeds was not assured. Once planted in moist ground, the seed crops might fail to germinate, or the seedlings might fail to flourish, or mature plants might fail to produce edible parts at the end of the growing season. But seed trials were comparatively cheap and easy to continue either until success was met or the project was judiciously abandoned.

In contrast, plantain introduction was complicated and expensive. As the European traders were untutored in the crop's biology, they first needed to recruit knowledgeable farmers who were familiar with the crop. Because the plantains reproduced vegetally (that is, new plants grew from active, living parts of parent plants in the form of 'suckers' that spread from the base of the parent), transferred samples were fragile and perishable.[47] Plantain introduction required expert hands to carefully excavate soil from around the suckers growing at the base of mature plants; to transfer these shoots into portable containers with packing materials suited to holding moisture; and to attend vigilantly to the samples to keep them adequately watered during their overland carriage, and then

44 Crone, trans. and ed., *The Voyages of Cadamosto.*

45 Denis Escudier, ed., *Voyage d'Eustache Delafosse sur la côte Guinée au Portugal et en Espange, 1479–81* (Paris: Editions Chandeigne, 1992).

46 G. H. T. Kimble, trans. and ed., *Esmeraldo de situ orbis* (London: The Hakluyt Society, 1937).

47 Rossel, *Taxonomic-Linguistic Study of Plantain in Africa, passim*; F. R. Irvine, *West African Crops* (Oxford: Oxford University Press, 1974), 86–90.

during their marine transport to the target areas of experimentation. During this part, substitution of saltwater for scarce shipboard freshwater would have immediately killed the suckers. Even if transported suckers arrived alive, they still might fail to grow or fruit if planted in inappropriate alien environments. But unlike seedy crops, plantain samples from faraway places were a relatively expensive and difficult to obtain genetic resource with which to experiment repeatedly.

Portuguese mariners used ships with sails to bridge the transportation gap, and they compensated for their own lack of practical knowledge about the plant by using expert Central African farmers to transplant this starchy crop to the Gold Coast. These expert farmers were slaves from the moist forests near the Central African kingdom of Kongo, where border wars of the newly centralizing polity created captives in the first decades of the sixteenth century. Portuguese traders transshipped these slaves via São Tomé, where a minority was sold to settlers then establishing sugar plantations, but most of the captive cargoes were delivered to the Gold Coast to trade against the gold the Portuguese primarily sought.[48] In these decades, Kongo displaced the coastal Niger Delta as the principal point of origin of slaves imported to the Gold Coast. These Kikongo-speaking slaves probably transplanted plantain suckers to the Gold Coast in the first decade of the sixteenth century, in the same era as their successful introduction of the crop to São Tomé[49] and other islands off the coast of Senegambia that Portuguese traders frequented.[50] The plantains they transplanted to these areas were of the False Horn bunch type, the group then predominant in West-Central Africa.[51]

Gold Coast farmers must have been impressed by the many good qualities of plantains as a food crop. They could prepare it into the same familiar foods as they did with yams, and the saw that the plantains reproduced vegetally like their familiar yams did, and thus it was especially easy to propagate preferred varieties. Also, plantains did not require the heavy labor of mounding earth and setting support stakes which yams needed.

[48] Miller, *Way of Death*, 115–16, 140–41.

[49] Valentim Fernandes noted that the crop had just begun to be cultivated there, circa 1506–10; T. Monod, A. Teixeira da Mota, and R. Mauny, trans. and eds., *Description de la côte occidentale d'Afrique (Sénégal au Cap de Monte, archipels) par Valentim Fernandes* (Bissau: Centro de Estudos da Guiné Portuguesa, 1951), 134–35, 188–90n294.

[50] By at least 1516, plantains were present on the Canary Islands, from which they were sent to the Caribbean island of Hispaniola; Rossel, *Taxonomic-Linguistic Study of Plantain in Africa*, 49.

[51] Rossel, *Taxonomic-Linguistic Study of Plantain in Africa*, 53–97, 183–88.

Significantly, farmers could harvest plantain fruits during the 'lean' months after most of the yam harvest had been eaten or else had spoiled.

Propagators selected a parent tree with desirable characteristic and selected a thick, 'maiden' sucker to cut off with a cutlass. They then replanted the cut sucker into ground they thought appropriate and had loosened up and covered the planted shoot with some organic material, such as mulch, to keep the sucker wet and to enhance the soil with nutrients. Once the sucker sprouted, propagators needed to weed around their plants so long as they were still small, which was a very brief period indeed for these vigorous growers. Twelve to fifteen months after planting, farmers harvested their first bunch and continued to produce bunches every four to five months. After four to five years, these plants tended to return fewer fruits to harvest and did so with increasing irregularity, but plantain farmers could choose to leave such plants standing for use as a cheap, labor-free buffer against famine.[52] Some plantain varieties produced fruits for a generation before finally succumbing to nematodes or other diseases and pests.[53]

Wives and daughters must have taken particular pleasure in the new crop, because when compared to yams or African grains, they found that a plantain was a raw ingredient that was easier for them to process into food. Plantain fruits were easier to peel than were yams, and didn't require the multiple stages of threshing and winnowing needed to process grains. Plantains also required less cooking, thereby reducing the labor inputs to collect firewood and monitor heat levels. Also, once cooked, the starch of most plantain varieties was easier to hand work into the desired final culinary form (especially the labor-intensive *fufu*).

Plantain farmers also discovered, or perhaps learned from slaves of the Central African diaspora, that they could harvest other domestic products from mature plants. They burned skins discarded while processing the fruits and the leaves from felled plants to yield ash very rich in potash and therefore very suitable for making soap, usually by mixing it with palm oil. Likewise, they stripped fruit stems and peduncles of their strong, potash-rich fibers, and used these to make frothy bath sponges and scouring pads. From longer fibers, they made fishing tackle and all-purpose twine.[54]

---

52 Irvine, *West African Crops*, 86–90.

53 Schoenbrun, *A Green Place, A Good Place*, 79.

54 Abbiw, *Useful Plants of Ghana*, 30; Rossel, *Taxonomic-Linguistic Study of Plantain in Africa*, 31–33; Irvine, *West African Crops*, 86–90.

Gold Coast farmers near the three Portuguese coastal outposts quickly incorporated plantains as an experimental crop to supplement their other starchy staples. Farmers of the coastal savannas near Elmina experimented with plantains in their kitchen gardens, but confronted difficulty in relocating the plants to new fields on the open grasslands. Although farmers discovered that plantains could tolerate saline air blowing onshore, strong winds absolutely shredded their plants' broad leaves and could topple a plant that had a hanging ripening bunch. Also, they found that although plantains did not require especially rich soils to produce fruits, their fruiting dramatically declined over the course of just a few years—roughly equivalent to grain crops they farmed. But unlike grain, for which a new fertile plot might be sowed each year or two, gardens of perennial plantain plants were not mobile. Also, plantains required moist ground but would fail if the soils were not well drained; and conversely, plantains on well-drained soils would succumb if rains were insufficient, which they often were on the extremely variable central and eastern coast.[55] Farmers here appraised plantains to offer significant additional risk and diminished cultivation labors only marginally. Therefore, they maintained plantains in kitchen gardens, where residential structures buffeted winds and where they could nourish plants with household wastewater and fertilizing rubbish.

Plantains were most important to farmers of the wet evergreen forests in the area of the outposts at Shama and Axim and in the moist deciduous woodland north of Elmina. Comparative linguistic evidence from these areas points to two separate paths of coastal introduction. Nzema-speaking farmers of the southwest borrowed their name for the plant directly from the Kikongo-speaking slaves in their midst and who introduced this crop. These slaves spoke of the plant they knew as *ebanga*, which designated a very popular False Horn cultivar grown in their natal communities. Nzema-speakers adopted this name as *bana*, and used it as a generic name for the experimental crop.[56] Fante-speaking farmers near Elmina named the crop using the neologism for 'overseas yam,' a name that speakers of other Akan dialects and southern Guang languages

---

[55] Ofori-Sarpong, "Rainfall Reliability over Ghana"; Papadokis, *Crop Ecologic Survey in West Africa*, maps 4, 5, 8.

[56] Gerda Rossel, "The Diffusion of Plantain (*Musa* spp. AAB) and Banana (*Musa* sp. AAA) in Africa: A Case for Linguists, Taxonomists, and Historians, focused on Nigerian Crop Names," in Richard E. Leakey and L. Jan Slikkerveer, eds., *Origins and Development of Agriculture in East Africa: The Ethnosystems Approach to the Study of Early Food Production in Kenya* (Ames: Iowa State University, 1991), 120–60.

borrowed when they began to experiment with the new crop. From these latter farmers, Ewe-speaking farmers on the eastern side of the Akuapem-Togo range acquired experimental suckers and the name for the crop, which they adapted into the phonology of their own speech.[57]

Forest farmers found plantains to be ideally adapted to the moist conditions of their gardens and their specific concerns. Moreover, because plantains, unlike yams, were adapted to areas without a dry season, forest farmers could pioneer all areas of the rainforests, including those especially wetter areas unsuitable for any other crop. In the evergreen forests of the southwest, farmers found that plantains planted on upland plots complemented their river-course rice gardens. But forest farmers everywhere needed to guard their plantain gardens against familiar pests, like monkeys that filched fruits just as they turned ripe and elephants that were known to take the entire plant.[58]

By 1572, plantains were well established in the crop strategies of forest farmers. In that year, a Portuguese visitor concluded that plantains were a wild part of the natural environment of the moist forests of the western Gold Coast. He reported that, 'on the road from Agri[59] I saw it [= a plantain plant] growing in the forests, where I think no-one could have planted it.'[60] But this foreign observer betrayed his ignorance of the plant. These plantains could not have been there without intentional human planting. Nor could they have been 'naturalized' in the horticultural sense, meaning that they were human-adapted plants that had escaped from gardens to reproduce on their own in the wild. Plantains cannot spread without human assistance. The plant he saw was probably a mature plant beyond its most fruitful years, perhaps standing on an old garden gone to fallow or one of its progeny by sucker. One type of plantain now grown in Ghana, the *borodewuio* (named for the color-resemblance of its pseudostem to that of the forest duiker; Akan, *owuio*) is known to stand on abandoned

---

[57] See Word List 1, page 75.

[58] For a report of raiding elephants in the late seventeenth-century, see Willem Bosman, *Nauwkeurige beschryving van de Guinese Goud- Tand- en Slave-Kust* (Utrecht, 1704), vol. 2, 24–25; his description was presented as if he personally sighted an elephant, but was probably based on hearsay, as elephants were probably not in the coastal area in which he resided. But the description is nonetheless a true representation of elephants' attraction to *Musa* gardens.

[59] Identified as Sumado (approximately 35 kilometers north-north east of Axim) in Kea, *Trade, Settlements, Politics in the Seventeenth-Century Gold Coast*, 70.

[60] This original document is in the Biblioteca National de Lisboa, Fundo Geral, MS 8457, ff. 100v-110. The text is also presented in transcription in Brásio, ed., *Monumenta Missionaria Africana*, sér. 1, vol. 3, 89–113. The text is presented in English translation with annotation in Hair, *East of Mina*, document E, pp. 73–94.

Word List 1. Comparison of Vernacular Names for Plantain

| Language | Dialect | Plantain | Comment |
|---|---|---|---|
| Akan | Fante | c. 1642 *broddÿ* (Annon. Dutch manuscript) | neologism, 'overseas yam,' *oboro* + *ode* |
| | | c. 1662 *broddi* (Müller) | |
| | | c. 1680 *obourady* (Barbot) | |
| | | borodze | sound change over time |
| | Nzema | bana | < Kikongo (H16) name for specific False Horn cultivar, *ebanga* |
| | Akuapem | c. 1783 *abrodeh* (Isert) | |
| | | borode | |
| | Asante | borode | 'overseas yam' |
| | | ɔbɔdeɛ (alt.) | |
| | Brong | bolodes | < Asante |
| Guang | Coastal | borode | < Akan |
| | Hill | borode | < Akan |
| Gã-Dangme | Gã | c. 1783 *amadah* (Isert) | |
| | | amadaa | independent innovation |
| | Dangme | manaa | |
| Gbe | Ewe | c. 1783 *abolodiu* (Isert) | < Akan or Guan (?via obsolete Gã term with common r > l sound change) |
| | | abladzu | sound change over time |

gardens and continue to bear fruit long after other plantain cultivars have succumbed to nematodes or other pests.[61] Farmers could use these sorts of plants as a cheap, if not terribly productive, food resource. They need not clear, plant, or weed—only harvest.

[61] S. K. Karikari, "A Note on Plantain (*Musa* AAB Group) and Banana (*Musa* ABB Group) Cultivars in Ghana," *Ghana Journal of Agricultural Science* 4 (1971), 81. Though long-yielding, this variety produces fruits which have large seed remnants (or, technically, aborted ovules) and its fruits remain relatively hard and 'dry' after boiling, therefore making them unsuitable for making *fufu*. Its fruits are however cooked and mashed with red palm oil into *edo*. [Note: the genomic formulations in the title and text of the article are consistently mistaken.]

Diminishing environmental moisture and highly unpredictable rainfall totals north of the deciduous forests and along the Accra plains constrained the frontier of Gold Coast farmers' experimentation with plantains. These were conditions similar to those of the Dahomey Gap that had previously restricted the westward passage of the crop from the Central African forests. Here, again, most dryland farmers abandoned experimentation with plantains, as they were unwilling to invest the enormous labor inputs necessary to reshape the landscape through irrigation and the planting of windbreaks. The persistence of a few Gã- and Dangme-speaking farmers in the Accra plains in establishing plantain gardens there, by locating scarce suitable sites which were available (notably urban spaces sufficiently removed from the coast), favored the development of generic names for plantains unrelated to those of their neighbors.[62]

The first wave of experimentation with False Horn plantains in the early sixteenth century was followed sometime before the end of the century by a second, this one focused on French bunch-types of *Musa* introduced from South Asia via West-Central Africa and São Tomé.[63] This new group included plantains that produced extremely heavy and compact bunches, as well as cultivars that produced sweet bananas which people could eat uncooked as a treat.

The frontier of experimentation with the new types followed familiar courses from coastal ports of entry to the forest heartland of False Horn gardens. Gold Coast plantain farmers probably approached the new sort of plantains with increased confidence, given their familiarity with its rustic relation, and thus French plantains probably found their place in principal gardens more quickly. Forest farmers valued French plantains as a prolific producer of fruit; indeed, farmers discovered that these plantains produced large and heavy bunches that required wooden supports to cradle the fruits and prevent the plant from breaking. But forest farmers found the labor required to grow this type to be a minimal investment towards a much greater reward. However, the new plantain never achieved the broad geographical expanse of the False Horn type, because French varieties thrived only within a narrow environmental range.

---

[62] The same naming phenomenon is found on the Bateke plateau (Central Africa), where farmers working this 'island' of dry, sandy soils just north of the Congo River developed generic names for plantains unrelated to those spoken in surrounding rainforests; Rossel, *Taxonomic-Linguistic Study of Plantain in Africa*, 111.

[63] Rossel, *Taxonomic-Linguistic Study of Plantain in Africa*, 94.

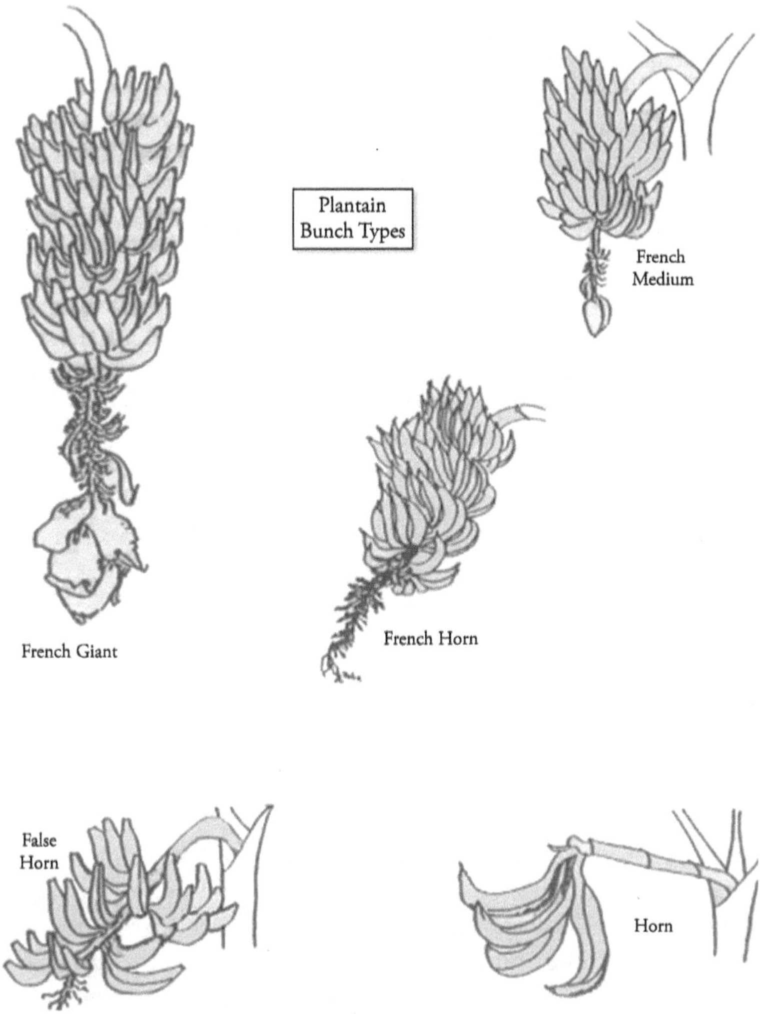


(source: adapted from Tezenas du Montcel 1987 via Rossel 1998)

Illustration 2. Plantain Bunch Types

Outside of the dense forest, plantain farmers preferred to continue to plant False Horn varieties, which they found to be hardier.

Gold Coast farmers never placed much importance on bananas. As most of this fruit's calories came in the form of simple sugars (fructose) rather than, as in plantains as complex carbohydrates, farmers recognized that the soft treats were unsuitable for processing into familiar starchy culinary forms. However, evidence from Gold Coast farmers' experimentation with bananas underlines again the critical role Central African slaves played in introducing *Musa* food plants to the Gold Coast and further demonstrates that earliest cultivation of plantains and bananas in this area belongs in the Atlantic-era of agricultural exchange.

Gold Coast banana farmers felt this crop to be sufficiently different from plantains to require a name of its own. Near the coastal ports, farmers borrowed names for bananas from Kikongo-speaking slaves, who spoke of bananas as *kondo,* the generic term for plantains in West-Central Africa.[64] Farmers on the Gold Coast borrowed this name, but they applied it specifically as their name for bananas. Atlantic-era word lists from the central Gold Coast show evidence of several separate introductions of banana cultivars before 1600 [see Word List 2; next page]. At that time, bananas were known in the coastal lingua franca under two names: *bachoven* and *bannana de Congo.* The later term clearly points to Central Africa as the regional source of the earliest bananas, and the crop was described as having been 'originally brought from the Congo Kingdom' by the Portuguese.[65] The term *bachoven* derived from the late sixteenth-century Portuguese word *pacova,* a term that Brazilian colonists borrowed from Tupi Amerindians' name for an indigenous plant (*Renealmia* spp.) that was of some physical similarity. Portuguese-speaking mariners who transshipped Brazilian sugar to northern European refineries introduced this name for bananas to Portuguese and Dutch traders then trading with West Africa. Neither term found lasting currency in any of the local Gold Coast languages but did indeed in other areas of early Atlantic commerce with Africa: on the island of Fernando Po and the nearby mainland; and in the Kru-speaking areas of Côte d'Ivoire to the west. West African slaves shipped to the plantation colonies of Suriname and Guyana continued to use this term

[64] The term *kondo* is a widely-shared cognate in most Western Bantu languages; Rossel, *Taxonomic-Linguistic Study of Plantain in Africa,* 50, 127–130; Rossel, "The Diffusion of Plantain (*Musa* spp. AAB) and Banana (*Musa* sp. AAA) in Africa." See also Vansina, *Paths in the Rainforests,* 61–65.

[65] Van Dantzig and Jones, trans. and eds., *Pieter de Marees,* 162–63, 166.

Word List 2. Comparison of Vernacular Names for Bananas

| Language | Dialect | Banana | Comment |
|---|---|---|---|
| Akan | Fante | c. 1642 *santamme* (Anonymous Dutch MS) | ?origin name for cultivar (AAB 'Pome') made generic term |
| | | c. 1662 *baccofes* (Müller) | coastal pidgin, *bachoven* |
| | | c. 1662 'a plant from São Tomé' (Müller) | see c. 1642 term |
| | | c. 1784 *santam* (Labarthe) | see c. 1642 term |
| | | mpua | < *mpantu* (a type listed in Christaller in the late 19thC) now named *apantu kwadu*, also known as *aborofo kwadu*; cultivar name for AAA 'Gros Michel' made generic term |
| | Nzema | kondo<br>konro (alt.) | < Kikongo (H16) |
| | Akuapem | kwadu | < Kikongo (H16) |
| | Asante | kwadu | < Kikongo (H16) |
| | Brong | kwado | < Kikongo (H16) |
| Guang | Coastal | kwadu | < Akan |
| | Hill | kwadu | < Akan |
| Gã-Dangme | Gã | c. 1858 *akwadu* (Zimmerman) | |
| | | akwadu | < Akan or Guan |
| Gbe | Ewe | c. 1783 *karatt* (Isert) | < ? |
| | | akɔdu | < Akan or Guan |

for bananas (now rendered *bak(o)be* in Papiamento and *bakove* in Sranan Tongo) and maintained *bana* as their name for plantains.[66]

[66] This linguistic connection was first noted by S. P. l'Honoré Naber, ed., *Beschryvinge ende historische verhael vant Gout Koninckrijck van Gunea* ('s-Gravenhage: Martinus Nijhoff, 1912), 167; and is described in more detail in Rossel, *Taxonomic-Linguistic Study of Plantain in Africa*, 125.

In the early decades of the seventeenth century, Fante-speaking farmers began to cultivate a new banana cultivar group brought to the Gold Coast from a proximate source in São Tomé. They named the new cultivar for its place of origin, as *santam*. This specific cultivar is difficult to identify botanically, but may refer to the AAB 'Pome' (or 'Prata'), a group of *Musa* cultivars that, despite their genomic formulation, produce sweet bananas that do not require cooking.[67] This banana was the first type to be introduced to Atlantic Africa.[68] This tentative identification also draws on similar names used by farmers of the moist forests of Central Africa lying north of the Congo River, who in the early seventeenth century were also in close contact with the Portuguese. They too named this cultivar group for its São Tomé origin.[69] The popularity of this cultivar on the seventeenth-century central coast favored Fante-speakers' use of the name for it as a generic label for bananas. In more recent times, this generic name has been replaced by Fante-speakers with a name that probably reflects the predominance of the AAA 'Gros Michel.' This cultivar is now known locally as *apantu kwadu*, or 'plantain-banana,' referring to the large, plantain-like size of the banana fruits.[70]

Plantain farmers recognized the new type of plantains under the generic name for the older group, evidently because they valued its similarity. However, forest farmers who cultivated both French and False Horn types of plantains developed a folk taxonomy to distinguish between the new plantains which produced more compact bunches with many individual fruits (French) and the older types that made less compact bunches with

---

[67] Refer to note 38 for genomic information. This cultivar is now known in the forest region as *alata kwadu*. In Akan speech, *alata* refers to Nigeria; laborers from Nigeria who migrated to Ghana during the 20th-century cocoa boom were said to favor this type; see Karikari, "A Note on Plantain (*Musa* AAB Group) and Banana (*Musa* ABB Group) Cultivars in Ghana," 83; and S. K. Karikari, "Some Taxonomic Assessment of *Musa acuminata* and *Musa balbisiana* to the Origins of Plantains and Bananas in Ghana," *Ghana Journal of Agricultural Science* 6 (1973), 16.

[68] Rossel, *Taxonomic-Linguistic Study of Plantain in Africa*, 94.

[69] These were Vili-speaking farmers of the Loango Coast and their inland neighbors; see Rossel, *Taxonomic-Linguistic Study of Plantain in Africa*, appendix D.

[70] Karikari, "A Note on Plantain (*Musa* AAB Group) and Banana (*Musa* ABB Group) Cultivars in Ghana," 83; and Karikari, "Some Taxonomic Assessment of *Musa acuminata* and *Musa balbisiana* to the Origins of Plantains and Bananas in Ghana," 16. I suspect this variety's alternative name, *aborofo kwadu* ('Europeans' banana'), describes its more modern economic importance as a crop grown for export to Western supermarkets rather than its introduction in early Atlantic times 'from overseas' (*oboro-*). Cf. *alata kwadu*, 'Nigerians' banana,' above. In the late 1950s, Ghanaian farmers who grew Gros Michel saw their plants succumb to Panama disease; in turn they increasingly adopted Cavendish for commercial farming, but now that variety is under siege. For a good, global history, see Dan Koeppel, *Banana: The Fate of the Fruit That Changed the World* (London: Penguin, 2008).

fewer, but larger, fruits (False Horn). They exaggerated the great number of individual fruits it produced, giving it the name *apem* (Akan, 'thousand') and classified the older type as *apantu* (Akan, 'large').

By the end of the sixteenth century, forest farmers' experimentation on the Gold Coast with this starch was complete. The new starch now contributed a significant part of the average diet of forest farmers. So central had plantain become to their food strategy that they already considered plantains to be an indigenous food crop. In 1600, the Dutchman Pieter de Marees heard, presumably from Africans residing near the central coastal areas he visited,[71] that plantains were among local crops that farmers there had 'always had.'[72] This historical vision is shared by many modern residents of southern Ghana.

The early consolidation of this historical understanding—initial local and oral, but has been folded in to the documentary record since De Marees's account of 1602—has skewed professional scholarship of the area's agricultural past ever since. The historiography takes no account of the etymology of local plantain names. Rather than reconciling the seemingly irreconcilable, scholars have preferred to follow *either* the linguistic clue of foreign origin *or* the oral tradition) of great local antiquity.[73] An examination of the structure of the non-debate demonstrates some of the special methodological challenges of doing research in a field like agricultural history that necessitates blending technical (botanical and agronomic) knowledge with historical methodologies, and marks many potential pitfalls for interdisciplinary research.

Scholars' choices between the two seemingly incompatible bases of evidence can be grouped by their respective academic disciplines. Botanists, foresters, and crop scientists working in Ghana have tended to follow the

---

[71] Just how De Marees managed to collect so much good original information about Elmina and its surroundings, despite avowed hatred between the Portuguese and the Dutchmen they considered to be pirates, remains a mystery. De Marees himself disclosed little about his sources, as his modern editors have averred the issue. Indeed, as one reviewer of the modern English translation of the work noted, this is the major lacunae in critical editions of De Marees; see J. D. Fage, "Seventeenth Century Gold Coast" [review of Van Dantzig and Jones, *Pieter de Marees*] *Journal of African History* 29, 2 (1988), 326–28. As Van Dantzig and Jones noted in their English edition, and before them Naber in his Dutch transcription, much, but not all, of what De Marees wrote about plantains was borrowed directly from Jan Huygen van Linschoten's 1596 account of Asia. See the modern editions: l'Honoré Naber, *Gout Koninckrijck van Gunea*, 167–69; and H. Kern, ed., *Het Itinerario van Jan Huygen van Linschoten, 1579–1592*, 2 vols., ('s-Gravenhage: Martinus Nijhoff, 1910), vol. 1, 226–29. See also the discussion of De Marees in Chouin, "Seen, Said, Deduced?"

[72] *'diemen hier... eertyts gheweest hebben,'* in Naber, ed., *Gout Koninckrick*, 173 (=Van Dantzig and Jones, trans. and eds., *Pieter de Marees*, 166).

[73] For a visual representation, see Illustration 3, page 89.

etymology and suggest an early Atlantic-era introduction of plantains to the Gold Coast. Academic historians and historically minded archaeologists, working in both Ghana and abroad, have generally argued for the great antiquity of *Musa* in the era, following contemporary oral tradition and the description of great antiquity as presented in De Marees's text.

The first historically minded scholar to wrestle with these contradictory strains of evidence were late nineteenth-century researchers involved with the Basel Evangelical Mission in the eastern Gold Coast. The Danish missionary-linguist Rev. J. G. Christaller described in his dictionary for Akan (which he called Twi) an etymology for *oborode* that indicated it to be an Atlantic-era introduction, and he listed several specific cultivar names.[74] After Christaller had seen his dictionary into print, he assisted in the editing and publication of a history of the Gold Coast written by another member of the Basel missionary community, Rev. Carl Christian Reindorf. The latter was a pioneering local historian of mixed racial and intellectual descent. His paternal grandfather was Danish and the other grandparents were African, and he drew on their distinct written and oral historical traditions to craft his reconstruction.[75]

Reindorf muddled the question as to whether plantains and bananas were indigenous by supplying no less than three different answers. First, he explained that the Portuguese had introduced both of these food crops, following the etymology of the plantain's Akan name as described by Christaller and, perhaps, an errant reading of the text of De Marees.[76]

---

[74] J. G. Christaller, *A Dictionary of the Asante and Fante Language, Called Tschi* (Basel: Basel Evangelical Missionary Society, 1881, 2nd ed. revised and enlarged, 1933). He listed the following types: *b(o)rɔde-fuferefu, b(o)rɔde-fufu, b(o)rɔde-hemma, b(o)rɔde-kokowa, b(o)rɔde-kwadu, obosim, agona-ne-toa, gyembum, nnweretia, ɔkɔm-bekum-wo, mpantu, mpemma, ɔsoboaso*. Most of these varieties are still grown in Ghana, and can be identified in the lists of cultivar names in Karikari, "A Note on Plantain (*Musa* AAB Group) and Banana (*Musa* ABB Group) Cultivars in Ghana," and Karikari, "Some Taxonomic Assessment of *Musa acuminata* and *Musa balbisiana* to the Origins of Plantains and Bananas in Ghana".

[75] Such a brief description as this fails to fully convey the intensely complicated problems of reading Reindorf. For a comprehensive analysis, see the seminar papers presented on the 100th anniversary of the publication of Reindorf's *History*, and now published as Paul Jenkins, ed., *Recovery of the West African Past: African Pastors and African History in the Nineteenth Century: C. C. Reindorf and Samuel Johnson* (Basel: Basler Afrika Bibliographien, 1998). Also Heinz Hauser-Renner, "Examining Text Sediments—Commending a Pioneer Historian as an 'African Herodotus': On the Making of the New Edition of C. C. Reindorf's *History of the Gold Coast and Asante*," *History in Africa* 35 (2008), 231–300.

[76] Reindorf may have also been aware of De Marees's book (if so, presumably the 1604 English translation) although he did not explicitly mention it among the written sources he consulted.

Second, he stated that there were oral traditions that described these crops as indigenous. And third, he gave what he called the 'true tradition' of how the plantain was discovered: a wandering hunter at Dompim [=?] in Akyem came across the unfamiliar fruit growing on the tree; the hunter was curious but cautious, and so first fed it to his dog to see if it was poisonous or not; when the dog survived, the hunter introduced the crop to human cultivation and consumption. To reconcile the opposing strains of evidence, Reindorf suggested a compromise. He gave an alternate, folk etymology for the Akan name for plantain as 'supplement to yam'[77] and further argued (in apparent contradiction to himself) that Christaller's etymology referred to 'different types' of plantains introduced by Europeans.

Botanists and plant scientists working in British colonial West Africa ignored Reindorf's complicated exposition and recognized Christaller's accent on Atlantic-era introduction as valid. J. M. Dalziel, a British botanist working in West Africa, noted that many vernacular names for *Musa* in the region indicated a foreign origin, and he too suggested a Portuguese-era introduction of this food crop to the region.[78] C. J. Taylor, a British forester working in newly independent Ghana, wrote that plantains had been introduced by Portuguese, following the etymology of the name first given by Christaller and the regional linguistic evidence from Dalziel.[79]

However, almost every historian and archaeologist interested in the Gold Coast past has looked past the linguistic evidence and the arguments of scientists who drew upon it. And they have written about plantains in a highly elliptical way that has made it difficult to discern when in the chronology of the area's agricultural past they placed original experimentation with plantains, and what weight they placed on which sources. But they have consistently implied that plantains were cultivated and consumed in the area before the Atlantic era. Perhaps these historians were initially hampered by the relative inaccessibility of the above-mentioned Portuguese sources, in which plantains were notably absent in West Africa before 1500. Compared to later Dutch-language travel accounts, which were not only published soon after their composition but also often saw

[77] Reindorf, *History of the Gold Coast and Asante*, 268. This etymology was so unacceptable to Christaller, who was acting as Reindorf's editor, that he felt it necessary to correct the proposed attribution in a footnote in the published edition.

[78] J. M. Dalziel, *The Useful Plants of West Tropical Africa* (London: Crown Agents, 1937), 468.

[79] C. J. Taylor, *Synecology and Silviculture in Ghana* (London: Thomas Nelson, 1959).

contemporary translation into English and French, few of the Portuguese manuscript sources became readily available to historians in widely distributed published editions before the mid-twentieth century.[80] But historians since then have not challenged the veracity of the oral tradition nor considered its historicity, despite a well-developed Africanist literature on the special problems of oral sources.[81] Nor have historians (beyond the translator-editors of the modern edition) typically interrogated De Marees's text with similarly sophisticated methodologies developed for analyzing written source material for pre-colonial Africa.[82] These deficits are more noticeable in light of the counter-evidence from Christaller's dictionary, which is widely used and respected by historians.

Ivor Wilks, the eminent historian of the pre-colonial Gold Coast and a formidable presence at Legon in the 1960s, authored a widely accepted description of farming in the pre-colonial forest region.[83] He related economic evolution from hunting and gathering to subsistence agriculture to the political development of the Asante state in the last years of the seventeenth century. His description suggested a form of 'traditional agriculture,' in which 'the factors of food crop production were constant,' and which, he stated, was substantially different from modern, cash cropping.[84] But to describe the significance of starchy staples in 'traditional agriculture' he drew consistently on quantitative data from modern farmers' carbohydrates, which he listed as African yams, plantains, cocoyams, maize, and cassava. Wilks did add some historical riders to his anachronistic reconstruction of an array of starches assembled mostly in the Atlantic era. He cautioned that maize's popularity had waned since the nineteenth century, and he reportedly adjusted some quantitative data to reflect the change from the original, African cocoyam (*Colocasia esculenta*) to the West Indian sort (*Xanthosoma mafaffa*), which became popular from the 1840s.[85] Later, he amended his reconstruction, 'I now believe that the full and decisive transformation

[80] Fage, *Original Sources for Precolonial Western Africa, passim.*

[81] Joseph C. Miller, ed., *The African Past Speaks: Essays on Oral Tradition and History* (Folkestone: Wm. Dawson and Sons, 1980); David Henige, *Oral Historiography* (London: Longman, 1982); Jan Vansina, *Oral Tradition as History* (Madison: University of Wisconsin Press, 1985).

[82] Jones, *Zur Quellenproblematik der Geschichte Westafrikas*; Beatrix Heintze and Adam Jones, eds., "European Sources for Sub-Saharan Africa before 1900: Use and Abuse," *Paideuma* 33 (1987) [special issue].

[83] Wilks, *Forests of Gold*, 40–90.

[84] Wilks, *Forests of Gold*, 41, 52.

[85] Wilks, *Forests of Gold*, 52, 53, 85n59.

to food crop cultivation in the Akan country probably had to await the adoption of high yielding plants of American origin: of the maize, cassava, and cocoyams that came to be the principal staples of the Akan farmer.'[86] What, then, was the historical place of plantains?

Wilks's description of adopting American crops significantly influenced many of his American and European colleagues. Raymond Kea argued that the introduction of new, highly productive starchy crops was of critical significance to changes in early political economy of the Gold Coast, but he did not describe plantains as one of these introduced crops, nor did he give them much attention in any context.[87] As historical background to a study rooted in nineteenth-century inland history, Thomas McCaskie based his description of pre-colonial forest farming on Wilks's account, with its implication that plantains were an ancient crop in the forest region.[88] Adam Jones, editor *par excellance* of primary source material for pre-colonial western Africa, translated early Akan names as they appeared in the primary sources he edited by supplying Christaller's etymology, thus indicating foreign origin. But the edition of De Marees co-edited by Albert van Dantzig, and which therefore must be understood as borne of compromise via collaboration, annotated De Marees's description of '*Bannanas* fruit, also known as Indian figs' and '*Bachovens*, also known as *Bannana de Congo*,' with the suggestion, '[i]t has been widely supposed that one of these was the plantain. Since neither De Marees nor any other seventeenth-century European visitor mentioned the cooking or frying of such fruits, it is doubtful whether the plantain had reached the West African coast by this time.'[89] It is perhaps problematic to argue from negative in De Marees's account of food because he, like most early European visitors who visited the coast and published their accounts, thought African food was inferior and generally had access to more familiar, European starchy foods by virtue of his privileged position. Only with the captivity of several Europeans in Asante in the later nineteenth century were good descriptions published about inland areas, and these notably included the making and eating plantain foods like *fufu*.[90]

---

86 Ivor Wilks, "'Slavery and Akan Origins?': A Reply," *Ethnohistory* 41, 4 (1994), 661.

87 Kea, *Trade, Settlements, Politics in the Seventeenth-Century Gold Coast.* Indeed, there are fewer than a half-dozen mentions of plantain in his entire monograph.

88 McCaskie, *State and Society in Pre-Colonial Asante*, 25–27.

89 Van Dantzig and Jones, trans. and eds., *Pieter de Marees*, 160n1.

90 See, for example, Claude-Hélène Perrot and Albert van Dantzig, eds., *Marie-Joseph Bonnat et les Ashanti, journal 1869–1874* (Paris: Musée de l'Homme/Société des Africanistes, 1994).

Ghanaian historians followed Reindorf's emphasis on ancient local origins more closely than have their Western counterparts. They have used his alternative etymology and repeated the oral tradition of the 'discovery' of the plantain found in his *History*. Perhaps they have also drawn on their own awareness of the same oral tradition, which is still told in southern Ghana.[91] Nana Akwasi Abayie Boaten argued that plantains were an ancient food of forest farmers. Abayie Boaten rejected the suggestion of a foreign origin for plantains given by Taylor (from the etymology first given in Christaller)[92] and repeated the etymology of indigeneity from Reindorf.[93] He also explained how plantains were discovered by a wandering hunter and his dog.[94]

Because remains of *Musa* plants do not survive in the archaeological record, archaeologists have tended to look towards historians' reconstructions for guidance on the antiquity of plantains and bananas on the Gold Coast. *Musa* plants which are used as food crops do not, in general, produce any physical evidence that archaeologists can find. They do not produce carbon-based organic substances such as wood or seeds, nor pollen;[95] nor do they produce phytoliths (telltale, microscopic mineral concretions of silica—the 'gallstones' of plants—which, because they are inorganic, do not decay as other plant parts)[96] that are sufficiently distinct

---

91 I was told the same oral tradition (twice) during fieldwork in Ghana in 1995. Does the modern tradition contain 'feedback' information from Reindorf? On the phenomenon, see David Henige, "The Problem of Feedback in Oral Tradition: Four Examples from the Fante Coastlands," *Journal of African History* 14, 2 (1973), 223–35.

92 'The assertion by C. J. Taylor that plantain was introduced by the Portuguese is very doubtful [in his] ref. *Synecology and Silviculture in Ghana*. Its name *boade* does not mean whiteman's yam'; in Akwasi Abayie Boaten, "Asante: the Perception and the Utilization of the Environment before the Twentieth Century," *Research Review*, new series, 6, 2 (1990), 88n5.

93 He suggested 'helper of yam' in Akwasi Abayie Boaten, "Kumase: Early Settlement up to the End of the 18th Century," *Research Review*, supplement 5 (1990), 13n10; and 'supplement to yam' in his dissertation: (née) Akwasi Boakye-Boaten, "Cultural Ecology of Asante, 1702–1945" (unpublished PhD diss., University of Ghana, 1974). 130. In the dissertation (pp. 130–31) it is apparent that he had read Karikari, "A Note on Plantain (*Musa* AAB Group) and Banana (*Musa* ABB Group) Cultivars in Ghana," because he repeated the same mistaken *Musa* genomic formulas (see above, note 61).

94 Abayie Boaten, "Asante: the Perception and the Utilization of the Environment," 20, 28n2. The tradition is also in the undergraduate thesis, K. Boaten, "A Historical Geography of Kwabere" (unpublished BA thesis, University of Ghana, [n.d.]).

95 A handful of *Musa* cultivars in East Africa which are used for fiber production and, in desperate times as famine foods, do produce pollen and (sterile) seeds.

96 The best modern survey of the technique is Delores R. Piperno, *Phytolith Analysis: An Archaeological and Geological Perspective* (San Diego: Academic Press, 1988). The technique is highly developed for grasses and trees, but identification criteria for most other plants, including *Musa*, is yet unproven.

from *Musa*'s indigenous, non-fruiting relation (*Ensete* spp.) to warrant confident identification.[97]

Archaeologists' approaches to plantains have therefore been in many ways similar to their neglect of yams, which also do not produce conventional archaeological remains.[98] Both starches are at the center of modern foodways in southern Ghana, and modern Ghanaians speak of both as indigenous. But whereas African yams have identifiable wild ancestors in the region, *Musa* has none and therefore cannot be considered to be native. In light of this, and perhaps because archaeologists are aware of the foreign indication in vernacular names for plantain, none have directly suggested that plantains were cultivated and consumed in pre-Atlantic times, as they have indeed argued for yams. As a result, archaeologists have not included plantains in their descriptions of agricultural change. Most noted the significance of plantains among the starchy staples now consumed in Ghana, as Wilks had done, but plantains have been also conspicuously absent from their lists of crops introduced in the Atlantic era as put forward by many of the principal archaeologists in this field, including James Anquandah,[99] Chris DeCorse,[100] and Ann Stahl.[101]

---

[97] Rossel, *Taxonomic-Linguistic Study of Plantain in Africa*, 5. Cf. the potential of the method Christophe Mbida Mindzie, *et al.*, "First archaeological evidence of banana cultivation in central Africa during the third millennium before present," *Vegetation History and Archaeobotany* 10 (2001), 1–6; and previously reported in personal communication from the Musée Royale de l'Afrique (Belgium) in Schoenbrun, *A Green Place, A Good Place*, 89n53. The academic controversy continues in: Jan Vansina, "Bananas in Cameroon c. 500 BCE? Not Proven," *Azania* 38 (2003), 174–76; C. M. Minzie, *et al.*, "Yes, There Were Bananas in Cameroon More than 2000 Years Ago, "*infoMusa: The International Journal of Banana and Plantain* 13, 1 (2004), 40–42; C. M. Minzie, *et al.*, "The initial history of bananas in Africa. A reply to Vansina," *Azania* 40 (2005), 128–35; Katharina Neumanm and Elizabeth Hildebrand, "Early Bananas in Africa: The State of the Art," *Ethnobotany Research and Applications* 7 (2009), 353–62; Roger Blench, "Bananas and Plantains in Africa: Reinterpreting the Evidence," *Ethnobotany Research and Applications* 7 (2009), 363–80.

[98] On the archaeological problems of recovering yam remains, see chapter two.

[99] Anquandah's list included maize and cassava from the Americas, and new types of yams, cocoyams, and rice from Asia; Anquandah *Rediscovering Ghana's Past*, 37–38, 130–31.

[100] De Corse's long list of crop plants included the starchy staples maize and cassava; Christopher DeCorse, "Culture Contact, Continuity, and Change on the Gold Coast, AD 1400–1900," *African Archaeological Review* 10 (1992), 187–89. The Van Dantzig and Jones edition of De Marees was an important source for DeCorse, as was personal communication with editors.

[101] Stahl listed only the New World carbohydrates, maize and cassava, and targeted them as a future research priority; Stahl, "Innovations, Diffusion, and Culture Contact: The Holocene Archaeology of Ghana," 103. Among her sources, Stahl cited Wilks, "Land, Labor, Capital."

The only direct comment from an archaeologist about the history of plantains in the area came from Merrick Posnansky, while he was head of the department at the University of Ghana, in Legon. In a book review in a little-circulated Nigerian journal, he wrote, 'The bananas of the Indonesians [=plantains] seem only to have reached West Africa relatively late in its agricultural history. West of Gabon the number of varieties of banana decrease rapidly and those along the coast were probably brought by European seafarers after 1500 AD.'[102] Posnansky referenced this information in a footnote to 'personal information from Mr. M. Gill, formerly of Department of Crop Science, University of Ghana.'[103] Michael M. Gill was an Israeli agronomist who was a temporary lecturer in Crop Science at Legon during Posnansky's tenure at the university. Gill published a brief paper a few years later that described several Ghanaian plantain cultivars, and there he cited Christaller and Dalziel among his sources. These were presumably one source of the personal opinion he communicated to Posnansky in 1969.[104] Posnansky's public, albeit relatively quiet, suggestion that plantains were a new crop to Atlantic-era farmers of the Gold Coast went unheeded or unheard, even by his own students.[105]

Thus the study of the early history of plantains on the Gold Coast has been something of a confusing round-robin of problematic interdisciplinary borrowing. Plant scientists listened to a missionary linguist's proposed etymology. Ghanaian historians relied on oral traditions and invented a folk etymology of their own to complement the oral history. Western historians have written ahistorically and thus avoided explicit acceptance of the folk etymology and the oral tradition, as well as the original linguistic evidence suggesting a recent, foreign introduction. But at the same time they have too often failed to follow the sources and thus left an implication that plantains were not an introduced crop, which is understandable given their focus on other sorts of history in which agriculture is a supporting rather than determining element. Archaeologists have relied on the models

---

102 Merrick Posnansky, "Review Article: Yams and the Origins of West African Agriculture [review of D. G. Coursey, *Yams* (London: Longmans, 1967)]," *Odu*, new series, 1 (1969), 106.

103 Posnansky, "Review Article: Yams and the Origins of West African Agriculture," 107n12.

104 Michael M. Gill, "Annotated List of Plantain Cultivars in Ghana," *The Ghana Farmer* 15, 2 (1971), 88–93. Oddly, nowhere in his own article did repeat his opinion that plantains were an Atlantic-era introduction.

105 Including Anquandah and DeCorse. Akwasi Boaten was aware of Posnansky's statement, but chose to follow Reindorf; see Akwasi Boaten, "Kumase: "Early Settlement up to the End of the 18th Century," 13n9.

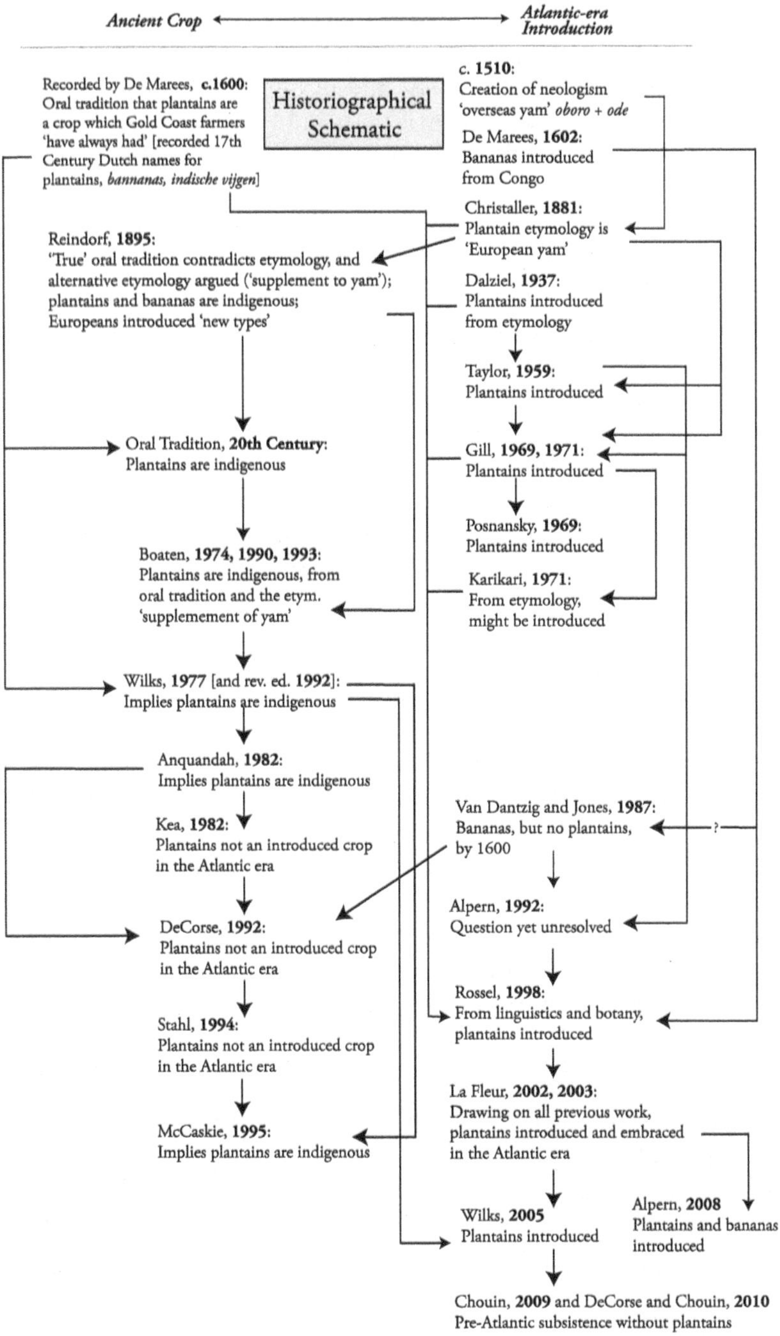


Illustration 3. Historical Schema of *Musa* Historiography

proposed by historians rather than pushing their own methodologies on a difficult part of their discipline, in which plantains are implied to have been a part of Gold Coast's ancient past and not to have been an introduced crop. It is no wonder that Stanley Alpern, in his comprehensive historical catalog of crop plants introduced to Western Africa in the Atlantic era, could write only that the question about the antiquity of *Musa* types in West Africa was full of contradictory evidence and thus 'remains open.'[106]

Comparative historical linguistic evidence from Gold Coast vernacular plantain and banana names show that *Musa* was indeed introduced from Central Africa. The environmental separation of the Gold Coast from Central African rainforests gives good reason for *Musa*'s absence in the region at the time of the documentary evidence from the era of early Portuguese reconnaissance of the area in the late fifteenth century. Documents from later eras of Atlantic trade describe the specific historical context in which Europeans were motivated to organize the introduction of this starchy crop. Written descriptions of *Musa* from the era of intensive trade point to several waves of introduction of different cultivars, and this is confirmed by the linguistic data.

### *Maize*

After 1500, farmers of the coastal savannas near Elmina began to experiment with maize (*Zea mays* L.) cultivation.[107] Presumably, they began their experiments with seeds they had acquired from the Portuguese outpost, which the farmers then planted in ways similar to their handling of native grains. They planted the seeds in advance of the long rainy season; then guarded them and monitored their progress through the season of growth; and then harvested their crop once the rains had faded and the grains were fully grown. Processors then dried maize ears in ways similar to millet

[106] Alpern, "The European Introduction of Crops to West Africa," 19–20. Alpern was familiar with many early Portuguese sources, Christaller, Dalziel, and the modern edition of De Marees. Alpern's updated survey is revised to show the history argued here; see Alpern, "Exotic Plants of Western Africa."

[107] The term 'maize' rather than 'corn' is used here to avoid unnecessary confusion. The identification of 'corn' with maize is typical to the (Anglophone) United States. However, for example, speakers of British English use the term 'corn' to describe whatever grain is most important to their local communities, and thus the term identifies different species of grain in different locales. The same is true among other Germanic-language communities (Dutch, Danish, Swedish, German) who share this cognate from the early Germanic *koren*, generic 'grain.'

and sorghum grain heads, by air-drying hanging bunches or by sun-drying grain ears on cleared, hard ground. Finally, cooks could substitute maize for the older grains in all of their recipes, including porridges made from cracked grain boiled in water, sour 'breads' made from fermented grain dough, and alcoholic and non-alcoholic brewed beverages.[108] The parallel cultivation, processing, and cooking techniques accelerated Gold Coast farmers' acceptance of maize as an experimental crop, and they encountered few environmental problems growing trial plots of this tropically adapted plant in Gold Coast environments. The age of early experimentation with maize on the Gold Coast was therefore brief, and its initial incorporation as a supplementary starchy crop was widespread.

Identifying precise dates for these early experiments is, as with all other agricultural innovations, impossible, but the earliest maize in this area (and the rest of the Old World, of Africa and Eurasia) is certainly post-Columbian. Only a minority of eccentric 'diffusionist' historians has proposed an ancient transfer of maize and other agricultural technologies via purported earlier trans-oceanic contacts.[109] Such arguments adventitiously marshaled all manner of selective and unsystematic linguistic evidence[110] and hasty conclusions from unfinished archaeological reports to suggest an ancient sort of African maize. However, there is no good evidence of maize cultivation on the Gold Coast before the early sixteenth century,

---

108 Abbiw, *Useful Plants of Ghana*, 23–24.

109 Notably, M. D. W. Jeffreys, "Pre-Columbian Maize in Africa," *Nature* 172 (1953), 25–35; and M. D. W. Jeffreys, "How Ancient is West African Maize?," *Africa* (International African Institute, London) 33 (1963), 115–31. For further comment on the controversy, see Alpern, "The European Introduction of Crops to West Africa," 24–25. Jeffreys also argued that Africans had made voyages to the Americas before Columbus; see M. D. W. Jeffreys, "Pre-Columbian Negroes in America," *Scientia* 58 (1953), 202–18, and supp. 113–28.

110 Most of these provocative arguments and subsequent opposing counter-arguments centered on which precise crops were identified in those sixteenth-century Portuguese sources which used variants of the lusophone term *milho*. See M. D. W. Jeffreys, "The Origin of the Portuguese Word Zaburro as Their Name for Maize," *Bulletin de l'Institut Française d'Afrique Noire* 29, ser. B, nos. 1–2 (1957), 111–36; and A. Teixeira da Mota and António Carreira, "*Milho Zaburro* and *Milho Maçaroca* in Guinea and in the Islands of Cabo Verde," *Africa* (International African Institute, London) 36, 1 (1966), 73–84; and Joaquim Lino da Silva, "O *Zea mays* e os milhos africanos na costa oriental da África: discussão em torno de documentos antigos," *Garcia de Orta, Série de Geografia* 14, nos. 1–2 (1993), 15–40. For general comment, see Alpern, "The European Introduction of Crops to West Africa," 24–25. This Portuguese term was later borrowed by Dutch (and, in some cases, Danish, English, and other foreign) traders on West African coast as *millie* (with various spellings), which they sometimes used with adjectives (albeit inconsistently), to identify Gold Coast grains. The seventeenth-century Dutch term *millie*, which by then almost exclusively indicated maize in references to western Africa, finds clear echoes in the speech of many modern South African languages as *mealies*, i.e., maize.

when Portuguese traders in West Africa introduced the first experimental seeds.[111]

The Portuguese did not value maize as a grain fit for human consumption, however they did think it appropriate for feeding livestock—or slaves. Portuguese residing in Europe first became acquainted with maize in the wake of Columbus's return from his first voyage to the New World. There, probably from Carib Amerindians of the island of Hispaniola (modern Haïti and the Dominican Republic), his expedition procured samples of local maize types. Following the return of Columbus to Europe in 1493, Iberian farmers began to experiment with maize cultivation. After 1500, they expanded their trials to include new varieties of maize brought from the new Portuguese claim of Brazil. Iberian farmers discovered that seeds from equatorial South America grew into plants that produced grains softer and easier to mill than the Caribbean types.[112] Portuguese consumers were suspicious of the new grain and endowed it with less prestige than their favored wheat. They also placed maize beneath lesser-esteemed European winter grains (rye, barley, oats), which made dark flour, and even beneath the 'rough' grains of millet and sorghum. They slandered maize as indigestible and, in the contemporary Portuguese insult, as only 'fit for pigs.'[113]

The Fante-speaking farmers near Elmina, who were probably the first Gold Coasters to experiment with maize, welcomed this crop without prejudice as an abundant supplier of grain. They drew a relationship between their native 'corn' (Fante, *wi*, 'millet') and this crop from overseas when they named it *oborowi*. Early experimenters must have been impressed with maize, because each maize plant could produce several ears of grain whereas familiar counterparts produced only a single cluster. Also, whereas grain farmers normally had to dispatch their children to the fields to protect exposed grains of sorghum and millet heads by scaring off birds which would otherwise prey on them, maize kernels were covered

[111] There is an archaeological claim of pre-Columbian maize in the northern pre-colonial Gold Coast, at a c. 500 BC settlement located along the White Volta, but the identification of maize is as weak as the chronological context is suspect; see Oliver Davies, "The Ntereso Culture in Ghana," in B. K. Schwartz and R. A. Dumett, eds., *West African Culture Dynamics* (The Hague: Mouton, 1980), 220.

[112] Irvine, *West African Crops*, 127; Frank Willet, "The Introduction of Maize into West Africa: An Assessment of Recent Evidence," *Africa* (International African Institute London) 32, 1 (1962), 11.

[113] Marvin P. Miracle, "The Introduction and Spread of Maize in Africa," *Journal of African History* 6, 1 (1965), 43n7. The same idiom appeared in late sixteenth-century descriptions of African cocoyams on the western Gold Coast; see Hair, *East of Mina*, 81.

by a protective sheath of papery leaves and therefore didn't need to be so closely monitored.

Coastal farmers' early-sixteenth century experiments with maize produced local varieties adapted to their specific environmental conditions. These farmers were particularly successful in developing two distinct types. The first was high-yielding maize grown during the longer 'big' wet season, falling in the months of April to June. Their second type was fast-growing strain known now in Akan as *apese buro*, which achieved a shorter terminal stalk height but returned a modest amount of grain at the end of the 'minor' wet season of mid-September to November (known in Akan as *ɔbέsέ*, hence the adjective for the maize as the *apese* type).[114] They also developed varieties of maize suited to particular foods: hard, floury maize for breads, and softer, moister maize to eat off the cob. To broaden the biannual maize harvesting season and avert failed crops, coastal farmers planted earlier-to-later ripening types in gardens of varying soil and moisture qualities. Early European visitors noticed maize ears with many different colors and combinations of colors, giving us some idea of another aspect of the wide genetic variety of early Gold Coast maize: '[t]hey are of various colors, such as white, black, yellow, purple, etc., and sometimes one sees no less than three or four different colors on one ear.'[115]

Farmers outside the area of the central coast quickly acquired seed specimens with which they began their own experiments. Recently published archeological findings show just how quickly farmers throughout the region began to experiment: by the middle of the sixteenth-century—that is at most just roughly fifty years after the earliest possible introduction of maize to the coastal region—farmers some 350 kilometers inland, in the moist savanna-woodlands of the Banda Hills, were growing maize.[116] The speedy embrace of maize throughout the Gold Coast favored a relatively uniform borrowing of Fante names for the experimental crop [see Word List 3].

In the Banda Hills, as on the central coast, farmers found maize to offer profoundly beneficial characteristics. In such moist environments, farmers

[114] I suspect this is what De Marees meant when he wrote, 'There are two types—large and small. The large maize is stronger than the small one'; Van Dantzig and Jones, eds., Pieter de Marees, 114 (= Naber, ed., *Gout Koninkrijck*, 119).

[115] Van Dantzig and Jones, trans. eds., *Pieter de Marees*, 114 (= Naber, ed., *Gout Koninkrijck*, 119).

[116] This is shown in the analysis of phytoliths reported (in an admittedly preliminary manner) in Ann B. Stahl, "The Archaeology of Global Encounters Viewed from Banda, Ghana," *African Archaeological Review* 16, 1 (1999), 5–12.

Word List 3. Comparison of Vernacular Names for Maize

| Language | Dialect | Maize | Comment |
|---|---|---|---|
| Akan | Fante | c. 1579 *bruy* (Hair, ed.)<br>c. 1662 *abruì* (Müller)<br>c. 1680 *abbrouann* (Barbot)<br>eburow | neologism: *oboro* + *wi* 'overseas millet' |
| | Nzema | abɛlɛ | |
| | Akuapem | c. 1783 *abró* (Isert)<br>aburow | <br>< Fante |
| | Asante<br>Brong | aburo | < Fante |
| Guang | Coastal | ablou | < Fante |
| | Hill | buruyo<br>kibwai | < Akan<br>< ? |
| Gã-Dangme | Gã | c. 1783 *ablé* (Isert)<br>abele | < Akan, with /r/ > /l/ change |
| Gbe | Ewe | c. 1783 *blofoë* (Isert)<br>bli<br>kpeli | <br>< Gã<br> |

could plant twice a year; although they discovered that maize flourished best on virgin ground, their bush clearing involved little more than cutting and burning short bushy undergrowth. Farmers in more heavily wooded environments between the coast and the Banda Hills, encountered more challenging clearing requirements. Forest farmers learned that maize quickly exhausted cleared soils, and that they would have to consistently organize substantial labor cooperation if they wished to carve new maize fields from the closed forest year after year. These farmers also experienced problems storing their harvests of experimental maize, because moist conditions during the processing and storage of maize grain favored the growth of fungi that spoiled hard-won stores. These farmers tended to prefer sowing 'floury' types of maize, to be harvested fresh while the kernels were still full of moisture, and then roast them as a snack or a vegetable accompaniment to a main meal.

If the principal rains came later than expected, or were diminished throughout the growing season, maize plants would fail to produce edible ears. But when rains were sufficient, they discovered maize to be an outstanding producer of bulk grain, and they encountered few of the storage problems that robbed farmers of moister climes of their harvest. Toward the irregularly watered Accra Plains, Gã-speaking farmers located few garden patches that were sufficiently watered to sustain a single crop of maize. And although they found that virgin patches were comparatively easy to clear with fire, given the light bush cover there, these soils were also the least rich in nutrients; in such areas, early techniques of growing maize probably produced much foliage but little grain, thus making the crop more suitable for fodder than human food.[117]

The source of the earliest experimental maize seeds sowed by Northern Guang speakers on gardens north of the Volta River is difficult to determine. The words they used to name the new crop bear little relation to those used by their southern neighbors to name maize introduced via the Atlantic coast. It is possible that dryland farmers of the northern savannas first obtained maize seed and borrowed names for this crop from their sources, via northern, trans-Saharan contacts. A northern origin was first proposed by Roland Portères in the 1950s, but his argument inferred ancient routes of diffusion from the distribution of types of maize (flint vs. flour) in modern Africa;[118] this historical argument was problematic because it assumed no change in West African farmers' preferences for different types of maize.[119] However, recent comparative historical-linguistic evidence from northern Nigeria suggests that Portères's argument (if not his evidence) may best describe early experimentation with maize on the northern Nigerian savannas and, by inference, also the dry northern grasslands of the pre-colonial Gold Coast.[120]

### *Asian Rice*

Of all the newly introduced starchy crops, the early history of Asian rice cultivation on the Gold Coast is the most difficult to reconstruct. This difficulty arises principally from the close relationship between African rice

117 Papadokis, *Crop Ecologic Survey in West Africa*, map 17.

118 R. Poetères, "L'introduction du maïs en Afrique," *Journal d'agriculture tropicale et botanique appliquée* 2, nos. 5–6 (1955), 221–31.

119 That is, perhaps one type of maize (flour or flint) displaced the other in more recent times; see Marvin Miracle, "Interpretation of Evidence on the Introduction of Maize into West Africa," *Africa* (International African Institute, London) 33, 1 (1963), 132–35.

120 Roger Blench, Kay Williamson, and Bruce Connell, "The Diffusion of Maize in Nigeria: A Historical and Linguistic Investigation," *Sprache und Geschichte in Afrika* 15, 1 (1994), 9–46.

(*Oryza glaberimma* Steud.) and its Asian cousin (*O. sativa* L.). Both species grow in moist soils and have similar vegetation, and their grains are prepared into food through similar techniques. Local rice farmers in early times certainly considered the new rice to be just another variant of their familiar staple, and thus they universally applied the generic name of their native crop (an apparently widely shared cognate in West African languages) to the introduced Asian species.[121]

Linguistic data therefore offer few clues about the geographic location, methods, and timing of the introduction and the subsequent adoption of the Asian species in early times by other communities who spoke different languages [see Word List 4; next page]. Nor does archaeological literature throw much light onto this history, as modern researchers have not concerned themselves with the early history of rice cropping, probably because rice of either species is not nearly so nutritionally significant in the diets of the area's modern inhabitants or culturally celebrated as it once was.[122]

Reconstructing the roles of African and Asian types in the region's history of rice cultivation therefore inevitably rests on incomplete and imperfect evidence. Much of the analysis turns on the contemporaneous, but flawed and sparse, observations of outsiders. Documentary evidence from Europeans who visited the coast in the pre-colonial era is skewed by these visitors' total ignorance of the existence of a native African species of rice. They, without exception, thought all rice was Asian in origin, and therefore they considered all rice grown in Africa to have come from seed stock originally introduced from the East.[123]

---

121 For the cognates, see Roland Portères, "Les appellations des céréales en Afrique IX," *Journal d'agriculture tropicale et de botanique appliquée* 6 (1959), 189–233. Application of generic names for the ancient African crop onto the new Asian type was common throughout the rice-growing region of western Africa. Like the rice farmers the Gold Coast, the Mende- and Temne-speaking farmers of Upper Guinea (modern Sierra Leone) applied their generic names for the native species to the introduced type; Norman E. Borlaug, *et al.*, *Lost Crops of Africa: Grains* (Washington DC: National Academy Press, 1996), 29; Carney, "African Rice in the Columbian Exchange," 386. Elsewhere in western Africa where rice was an entirely new crop, farmers borrowed or invented names for Asian rice. For example, coastal Niger Delta farmers borrowed from the Portuguese *arróz*; Alpern, "European Crops in Precolonial Africa," 21.

122 Perrot, "Semailles et moissons dans la région d'Assinie vers 1700".

123 For example, De Marees, 'This seed originally came from India, but has been carried all around the world so much that one now finds rice everywhere'; in Van Dantzig and Jones, trans. and eds., *Pieter de Marees, Description and Historical Account of the Gold Kingdom of Guinea*, 159. Also Barbot, '[rice] was first brought from India'; in P. E. H. Hair, Adam Jones, and Robin Law, trans. and eds., *Barbot on Guinea: The Writings of Jean Barbot on West Africa* (London: The Hakluyt Society, 1992), 197.

Word List 4. Comparison of Vernacular Names for Rice

| Language | Dialect | Rice | Comment |
|---|---|---|---|
| Akan | Fante | c. 1600: *amo* (De Marees)<br>c. 1662: *emu* (Müller)<br>emo | |
| | Nzema | emo | |
| | Akuapem | ɛmo | |
| | Asante | ɛmo | |
| | Brong | ɛmo | |
| Guang | Coastal | | |
| | Hill | imwe | |
| | Northern | a-malo | |
| Gã-Dangme | Gã | c. 1850 *ɔmo* (Zimmerman)<br>ɔmo | |
| Gbe | Ewe | mɔlu | |

Furthermore, the close physical similarity between the two species meant that even if visitors had been aware that there were two different types, few of them would have possessed the botanical knowledge to distinguish between plants of the two species.[124]

However, incidental comments in documentary sources about the qualities of rice grain seen for sale at coastal markets in the early Atlantic era suggest some tentative identifications. European writers unwittingly noted subtle visual differences between processed grains of the two species: while processed grains of the Asian species were a (to us, now familiar) brilliant white color, the African types had a red 'skin' of bran covering the grain's endosperm. Early European visitors to western Africa noticed color because they considered 'white rice' to be the most healthful and regarded other shades as being of inferior quality. From tentative identifications based on these early descriptions of color and subjective appraisals of 'quality,' a preliminary historical reconstruction of precolonial rice cultivation on the Gold Coast can be put forward.[125]

124 When compared to Asian rice, the African sort has smaller ligules (thin membranes found where the leaf joins the stem), has more compact panicles, and lacks awns (tiny bristles at the tip-end of spikelets).

125 I note that some early-published European accounts contain illustrations of certain crop plants, including rice, which might be used to help identify species, but I consider these illustrations to be of spurious historical significance. Rarely were these renderings

Before the sixteenth century, Gold Coast farmers were on the margins of the so-called rice belt stretching from Upper Guinea (modern Sierra Leone, Liberia, and coastal Côte d'Ivoire) to the inland Niger delta (modern Mali), where African rice was the principal staple of farming communities. On the Gold Coast, rice cultivation was localized in geographically distinct areas, each with its own distinctive rice-growing strategy. In the moist evergreen rainforests of the southwest, farmers sowed water-loving lowland types of African rice in marshy plots and along river valleys, and upland varieties on the highlands that framed the watercourses. These communities prized African rice as a highly productive starchy crop that required less labor to produce than equal volumes of cultivated yams, which supplemented their diet. Agriculturalists of the semi-deciduous forest south of the Kwahu Plateau and in the forested highland of the Akuapem-Togo Range planted African rice as an important supplement to the yams they preferred, and they made their rice gardens among poorly drained soils that would not support any other food crop.[126] Farmers who worked alluvial soils lying alongside rivers that fed the Volta used flood-advance rice planting techniques as a hedge against failure of their dryland crops, such as sorghum, millet, and some adapted varieties of yams.[127]

These Gold Coast rice harvesters found that the red bran of their African rice was very difficult to remove with the wooden mortars and pestles that they used to process the seeds. Once the husk was removed by striking and winnowing, processors could continue to work the grains in a mortar to remove much of the reddish bran that remained fixed to the seed, but success came at the expense of many broken grains, because the African types were brittle and broke apart easily. Because rice, unlike most other grains, was consumed in the form of whole grains rather than

created by the authors of the texts, but rather most often by commercial artists in Europe who were completely without any first-hand experience of their own with the subject they drew. Instead, the artists normally relied on the rough descriptions in the accompanying text, often combined with other, unacknowledged sources which concerned other 'exotic' locations; Jones, *Zur Quellenproblematik der Geschichte Westafrikas*, 99–105; Adam Jones, "Drink Deep, or Taste Not: Thoughts in the Use of Early European Records in the Study of African Material Culture," *History in Africa* 21 (1994), 349–70. Cf. Regula Iselin, "Reading Pictures: On the Value of the Copperplates in the *Beschryvinghe* of Pieter de Marees (1602) as Source Material for Ethnohistorical Research," *History in Africa* 21 (1994), 147–70.

[126] On farming in selected wetlands of Akyem: Abayie Boaten, "Asante: the Perception and Utilization of the Environment before the 20th Century," 25.

[127] Abbiw, *Useful Plants of Ghana*, 24; Boaten, "Asante: the Perception and Utilization of the Environment," 25; and for the Akuapem-Togo Range, see Borlaug, *et al.*, *Lost Crops of Africa*, 21.

used as ground flour, cooks and consumers found this sort of breakage not only unattractive but undesirable, because broken grains cooked unevenly and were difficult to eat with their hands. Nor was removing the red bran necessarily preferable, as this bran imparted a desired nutty flavor on the grain while cooking. Grains cooked with their bran intact were also more nutritious, particularly in thiamin. Thus even among the Gold Coast communities where rice was the principal starchy staple, consumers avoided thiamin deficiency disease, known as beriberi.[128] Although consumers were probably oblivious to the connection between the red bran and the inclusion of this important vitamin in their diets, they might have felt as some modern rice eaters do that red rice was more 'filling.'[129]

Sometime in the sixteenth century, Portuguese seamen brought grains of the Asian type to the West African coast. Introduction of the Asian seeds need not necessarily post-date Iberian exploration of the Far East in the late fifteenth century,[130] as Atlantic traders already had access to Asian rice seed grown in Western Europe. Farmers of the northern Mediterranean had been cultivating crops of the Asian species for many hundred years.[131] There is a claim that the Asian species had already been introduced across the Sahara before the Portuguese introduced their samples,[132] but medieval Arabic sources that describe Sudanic rice culture lack any information that would support such specific identification of the ancient crop as Asian.[133] In contrast, there is convincing linguistic evidence from

---

128 Beriberi (Sinhalese, reduplication of *beri*, 'weakness') afflicts the peripheral nervous system, and is characterized by partial paralysis of the extremities, emaciation, and anemia. Rice eaters of pre-industrial Asia, who preferred their rice cleaned of bran, dodged beriberi by accident of their labor-saving processing techniques; their method of parboiling husked rice so as to loosen the bran and thereby make it easier to remove had the effect of forcing thiamin from the nutrient-rich skin into the kernel. Not until the replacement of traditional techniques by mechanical milling in the nineteenth century did beriberi become epidemic in Asia.

129 Borlaug, *et al.*, *Lost Crops of Africa*, 29–30.

130 That is, when Vasco da Gama reached India in 1498.

131 Alexander the Great had introduced Asian rice to the Mediterranean around 300 BC, but not until the Moors planted large quantities in Spain in the eighth century was this crop widely cultivated in southern Europe (specifically Valencia and the Camargue); Andrew M. Watson, *Agricultural Innovation in the Early Islamic World* (Cambridge: Cambridge University Press, 1983), 17; Raymond E. Crist, "Rice Culture in Spain," *Scientific Monthly* 84, 1 (1957), 66–74; Thomas F. Glick, *Irrigation and Society in Medieval Valencia* (Cambridge: Cambridge University Press, 1970).

132 Raymond Mauny, *Tableau géographique de l'Ouest africain au moyen âge d'après les sources écrites, la tradition et l'archéologie* (Dakar: l'Institut français d'Afrique noire, 1961), 242.

133 Alpern, "European Crops in Precolonial Africa," 21.

the seaside, at the Niger River's Atlantic delta, that the Portuguese were indeed involved in introducing the Asian species.[134]

The Portuguese initially introduced Asian rice to the areas where they purchased slaves, in Upper Guinea and the coastal Niger Delta, as well as to areas they settled, as on the island of São Tomé and riverine trading settlements in Senegambia. The development of Asian rice cultivation in areas of Portuguese colonization allowed settlers to continue consuming a familiar starch from Europe, but one that flourished in western Africa. Slave traders hoped that farmers would develop Asian rice as a commercial crop, because processed rice, unlike the existing local starches (notably yams), stored very well and could be used to feed slave cargoes on the long voyage to Brazil.

It is impossible to date precisely when Gold Coast rice farmers first started to experiment with Asian rice, but their early experiments probably postdate those in areas of settlement or slave exporting. In the early sixteenth century, the Portuguese castle at Elmina was supplied with 'Guinea rice' from São Tomé,[135] and rice continued to be a significant food crop on the island at the mid-seventeenth century.[136] It is likely that farmers at the heart of the rice belt in Upper Guinea had begun to crop Asian types by 1574, when African marketers on the Sierra Leone coast were selling rice 'as fine as that of Valencia and very white,' which, it has been noted, suggests that this crop was indeed the Asian species.[137] But once the Asian species was introduced to the Gold Coast, local rice farmers did not abandon their native type of rice for the newly introduced one immediately. Rather, only once their experiments had shown them that Asian

---

[134] See note 118, above.

[135] Ballong-Wen-Mewuda, *São Jorge da Mina*, unpaginated chart ("Mouvement des navires de traite et des marchandises africaines"), facing p. 358.

[136] Adam Jones, trans. and ed., *West Africa in the Mid-Seventeenth Century: An Anonymous Dutch Manuscript* (Atlanta: ASA Press, 1995), 58.

[137] A. Teixeira da Mota, P. E. H. Hair, and L. Bourdon, trans. and eds., *A. Donelha, Descrição da Serra Leoa e dos rios de Guiné do Cabo Verde, 1625* (Lisbon: Centro de Estudos de Cartografia Antiga, Secção de Lisboa: 1977), 81, 211n48. A mid-seventeenth-century Dutch guide to West Africa noted that 'a large amount of gray rice' was available at Cape Mount (Upper Guinea) and that Dutch factors had previously been posted there to buy this rice as provisions; Jones, *An Anonymous Dutch Manuscript*, 21. Asian rice cleaned of its bran is white in color, but is susceptible to mildew and other fungi if harvested (or processed or stored) in moist conditions; see O. A. Danquah, "Fungi Associated with Discoloured Rice (*Oryza sativa*) Seeds in Ghana," *Ghana Journal of Agricultural Science* 9, 3 (Dec 1976), 185–87.

rice had desirable qualities did local farmers incorporate the new variety into their cropping strategies alongside the older type.

Farmers were initially interested in experimenting with the new type of rice to broaden variety in the seeds they sowed, so as to counter annual exposure to failed fields destroyed by pestilence or undesirable weather conditions. Later, they might have also recognized the commercial potential of supplying 'white rice' to Portuguese seafarers, who had originally introduced the crop and thereafter sought it for shipboard provisions. Experimenting farmers discovered that stems of the new rice were stronger, and therefore less likely to blow over in storms and become matted together, destroying the crop. Their strength was an especially attractive quality for the rice farmers of the southwest coast and northern plains, whose gardens were exposed to high winds at the change of the seasons, when the seed heads were nearly mature. Farmers recognized that the Asian rice was a highly developed crop that was less 'weedy' than African types that put more energy into the growth of inedible foliage at the cost of diminished seed size and number. Experimenters also found that the husks of the Asian type did not shatter to drop their seeds in the fields like the African types did. These farmers could therefore harvest more seeds per plant, and the new crop's retention of its seeds allowed harvesters to concentrate their collection efforts to when the last of the field was mature, without suffering significant losses to dropped, earlier-maturing grain.[138] They discovered that the Asian grain was harder and thus could withstand their processing techniques without significant breakage, and that they could thereby produce grain European buyers were prepared to purchase at higher prices.

These farmers also learned through trials that they could not easily cross-pollinate the two sorts of rice—not passively by planting them in close proximity or more actively by manually transferring the grasses' pollen—to create new hybrids with desired characteristics. This genetic isolation hampered the Asian types, because they were unable to quickly acquire any of the genes that endowed cultivated varieties of African rice with moderate resistance to native diseases and pests. Whereas African cultivars participated in the gene flow between planted fields and their wild cousins, from which they obtained such resistances as well as their

[138] Borlaug, *et al.*, *Lost Crops of Africa*, 17–37; Irvine, *West African Crops*, 130–35; Abbiw, *Useful Plants of Ghana*, 24.

undesirable tendencies toward weediness, the Asian species was more—but perhaps not completely—removed from this potentially enriching interaction.[139]

Farmers thus could not overcome some of the comparatively negative qualities of Asian rice through their selective breeding within the species. Asian rice demanded much more labor throughout the season, from sowing to harvesting, if the crop were to be successful. Sowers needed to plant their crop in closer relationship to the onset of seasonal rains; otherwise, the seeds would fail to sprout because residual ground moisture was sufficient to induce spoilage but inadequate to sustain sprouting. Then, farmers needed to weed their Asian rice gardens continually, because seedlings grew more slowly and were therefore less able to compete with weeds. Farmers could employ familiar irrigation techniques to suffocate weeds, but this required additional labor inputs to construct ridges and drains and then to monitor water levels, because the Asian types flourished within only a narrow range of water depth.[140]

Farmers of the southwestern river courses, who already had rice as their principal starch, were probably the first in the area to plant substantial quantities of the foreign species. For them, adopting the Asian type required only that they readapt existing techniques to the specific, but similar, requirements of the new sort. Biological differences in the two species, however, favored the development of segregated fields, to better manage water levels and to appropriately time sowing, weeding, and harvesting. From the sixteenth century, these farmers began to replace African rice with the Asian sort, but they never completely removed the old sort from their agricultural strategies during the Atlantic era. By 1600, women at the large produce market near Elmina were selling rice, which appears to have been of the Asian species.[141]

---

[139] Borlaug, *et al.*, *Lost Crops of Africa*, 17–37; Irvine, *West African Crops*, 130–35; Abbiw, *Useful Plants of Ghana*, 24. Since the identification African rice as a species distinct from the Asian type, it has been thought there is a sterility barrier that makes impossible genetic interaction outside of the laboratory. Emerging research suggests introgression of the Asian species's genes into some African types. See E. Nuijten, R. van Treuren, P. C. Struik, A. Mokuwa and F. Okry, *et al.*, "Evidence for the Emergence of New Rice Types of Interspecific Hybrid Origin in West African Farmers' Fields," *PLoS ONE* 4, 10 (2009). As the authors note, these interspecific hybrids are rare or absent in modern Ghana.

[140] Borlaug, *et al.*, *Lost Crops of Africa*, 17–37; Irvine, *West African Crops*, 130–35; Abbiw, *Useful Plants of Ghana*, 24.

[141] The rice was described as having a yellowish bran, and, '[w]hen taken out of these pointed husks, the grains are whiter, clearer, harder, and much nicer to make porridge

Contemporary Dutch reports suggest that farmers of the southwestern coast sold grain of both species into the eighteenth century[142] and that farmers of the eastern Gold Coast had begun to cultivate the Asian type by the late eighteenth century.[143] But these early crops of Asian rice, perhaps grown opportunistically to sell to Europeans, who relied on this market for local provisions, did not immediately replace fields of the older type.[144]

They continued to plant some African rice every year as part of their traditional tactic of limiting risk through diversity, and perhaps in especially lean times planted some quick-maturing African seeds to break the hunger season before core crops of Asian rice could be harvested. These farmers could also rely on African rice to shatter its seeds and thus sow itself on disused ground, then fend for itself against encroaching weeds and threat of disease, thereby providing a labor-efficient buffer against famine in bad times or, in periods of good harvests, supplemental food resources which freed surpluses of more valuable Asian rice to be sold.[145] Rice marketers perhaps tried to pass off grains of the African sort as if they were the more costly 'white' Asian type, by trying to remove the reddish bran. In the early nineteenth century, a Dutch administrator complained that the rice sold at Axim was of high-quality, and 'not inferior to

---

with than the best wheat that exists'; Van Dantzig and Jones, trans. and eds., *Pieter de Marees*, 158. These editors, however, suggested that this was likely to have been African rice.

142 Noting European distaste for the 'unsifted and unhusked' rice grown in the area was: Bosman, *Nauwkeurige beschryving van de Guinese Goud- Tand- en Slave-Kust*, vol. 2, 77. Bosman's work was the subject of a quick, but quite accurate for its time, English translation [1705; and then in facsimile, 1907 and 1967]. Albert van Dantzig improved the English translation and added some annotation, in his annual series: "English Bosman and Dutch Bosman: A Comparison of Texts," *History in Africa* 2 (1975) to 11 (1984). My translation of this particular passage is, however, different from Van Dantzig's, where the rice was described as 'husked' (i.e., with the husk removed) but was annotated as 'paddy,' a term which describes rice with its husk (or 'hull') still on; and the term also describes a method of irrigated cultivation used primarily for the Asian species (from Malay, *padi*, 'rice field'). Bosman was drawing a contrast between the 'perfectly cleaned' rice (that is, arguably, Asian rice) available in Upper Guinea with the 'unhusked' (that is, bran-on African rice) that was available on the western Gold Coast.

143 F. N. Hepper, *The West African Herbaria of Isert and Thonning: A Taxonomic Revision and Index to the IDC Microfiche* (Kew: Bentham-Maxon Trust and the Royal Botanical Gardens, 1976), 147. These Danish botanists recorded that Asian rice was 'cultivated,' presumably in the area of Akuapem, where they are known to have collected plant specimens.

144 Locating Asian rice cultivation on the Gold Coast in the Atlantic era rather than the colonial era is *contra* the assertion in Carney, "African Rice in the Columbian Exchange," 386–87.

145 Recording the importance of 'uncultivated' rice, i.e., self-shattering African types, is Quartey, "Development of Human Geography of Ahantaland before 1600," 124.

Carolina rice,' but came broken into small pieces, he thought because of the 'clumsy' manner in which Africans handled the grain.[146]

Once Gold Coast farmers had taken the full measure of Asian rice's particular advantages, they began to incorporate this new crop into their food gardens. They did not automatically embrace Asian rice as an inherently superior cultivar. Nor did European colonists in North America consider African rice to be inferior. The history of African rice in the Atlantic diaspora is also often overlooked, as literature on the so-called 'Columbian exchange' focuses on the movement of American and Asian crops into Africa but has often overlooked the importance of African cultigens, including their rice, in the early agriculture of the Americas.[147] That Asian rice is more productive than the African species is a modern notion,[148] and one which conceals the multiple historical considerations on which Gold Coast farmers based their decisions to plant the new rice—or not.

## *Conclusion*

The frontier of agricultural experimentation rippled unevenly throughout the area as local farmers tried new plants in their kitchen gardens, selected appropriate ones, and adapted them to their own preferences and the natural environmental limits of their garden plots. The era of early experimentation with foreign starches came to a close when farmers throughout the Gold Coast had tested the new crops, and based on their subjective assessments of the new crops' performance, had decided to include the new foods in their gardens in varying ways, or to reject them entirely. Of course, farmers who saw potential in the new crops continued stubbornly to try to perfect their techniques to maximize what favorable characteristics they perceived in the new starches, but by then the immediate benefits and inherent botanical limitations of the new starches were known throughout the area.

---

[146] Johannes Adam de Marrée, *Reizen op en beschrijving van de Goudkust van Guinea*, 2 vols. ('s-Gravenhage and Amsterdam, 1817–18), vol. 1, 93.

[147] Judith Carney, "The African Origins of Carolina Rice Culture," *Ecumene: A Journal of Cultural Geographies* 7, 2 (2000), 125–49; Judith Carney, "The Role of African Rice and Slaves in the History of Rice Cultivation in the Americas," *Human Ecology* 26 (1998), 525–45; and Carney, *Black Rice*.

[148] Cf. Alpern, "European Crops in Precolonial Africa," 20; Portères, "Vieilles agricultures de l'Afrique intertropicale," *L'Agronomie Tropicale* 5, (1950), 489–507.

In the era of early Atlantic contact, Gold Coast farmers displayed characteristic, and enduring, curiosity and openness to change.[149] Gold Coast élites and their subjects who worked the land pursued new agricultural technologies and technical resources.[150] In the early Atlantic era these farmers eagerly experimented with growing and eating foreign foods. They targeted their early experiments at producing a locally viable crop, that is one they could plant in their kitchen gardens and then see to produce edible parts. Local farmers pursued experimentation in hopes of finding cultigens that would supplement existing subsistence strategies so that they might limit risk of hunger as threatened by inter-year weather variability and crop-specific pestilence.

In contrast, European residents on the Gold Coast were quite rigidly, rather than adaptively, conservative in their food preferences. Initially they hoped to continue familiar foodways based on European carbohydrates. However, in the face of environmental limits on these crops, they experimented with starchy food crops they acquired elsewhere in their expanding worldwide empire. They introduced new starches from abroad to West Africa for their own purposes, if not initially for their own consumption. Only in the face of immediate need, such as palpable hunger, did the expatriates adopt previously unthinkable new starches in their diets.[151]

149 T. C. McCaskie, "Innovational Eclecticism: the Asante Empire and Europe in the Nineteenth Century," *Comparative Studies in Society and History* 14, 1 (1972), 30–45; D. Maier, "Nineteenth-Century Asante Medical Practices," *Comparative Studies in Society and History* 21, 1 (1979), 63–81.

150 Cf. the argument that African cultures systematically suppressed all but the most immediately necessary innovations; in Robert Horton, "African Traditional Thought and Western Science," *Africa* 37, 1–2 (1967), 51–71, 155–87.

151 Their Spanish colonial contemporaries in Central America followed a similar historical course down the metropolitan order of culinary preferences to eventually subsist on local starches; see J. C. Super, *Food, Conquest, and Colonization in Sixteenth-Century Mexico* (Albuquerque: University of New Mexico Press, 1988).

CHAPTER FOUR

# YOU REAP WHAT YOU SOW:[1] THE PROFITS AND PERILS OF THE NEW STARCHY STAPLES

Most Gold Coast farmers made either plantains or maize their primary starchy staple in the early Atlantic era, initially because they sought to integrate themselves into the Atlantic commercial world, and then later because they needed to survive the consequences of an intensifying engagement that they could not control. From the late fifteenth century, forest dwellers abandoned their previous dependence on yams in favor of plantains in order to feed the slaves and other dependants who extracted the gold craved by foreign traders in previously impossible amounts. On the coastal grasslands, initially nobles and then young entrepreneurs made themselves rich by exploiting the markets for dried maize grain presented by European buyers of slaves and slave shippers and then claimed cultural prestige for themselves by exalting their crop. Their good fortune contrasted with the hardships of their slaves and dependents whose monotonous diet based on leftover grain during eras of bountiful harvests left them rife with pellagra and, in times of drought and crop failure, made them vulnerable to impoverishment, pawning, enslavement, and Atlantic dispatch.

## *Going for Gold with Plantains*

In the early Atlantic era, local nobles and so-called 'big men' (Akan, *abirɛmpɔn*) who were wealthy but without the inherited prestige of royalty exploited the high productivity of plantain agriculture to expand their gold-mining operations in the forested country, and they used the high international commercial value of gold to aggrandize their wealth in locally understood terms of human dependents. This idea of enrichment was captured in the maxim, *Wo wɔ nipa a, na wo wɔ adee* (Akan, 'if you have people, you have wealth').

[1] Galatians 6:7: '[...] for whatsoever a man soweth, that shall he also reap.'

From the early fifteenth century—a hundred years before Gold Coast farmers first began to experiment with plantains—entrepreneurial-minded leaders probably extracted increased amounts of gold for export to the North, to trans-Saharan markets, and from the 1470s, to European traders on the coast in exchange for the slaves they primarily sought. Although efforts to analyze extant historical records of the gold trade are hobbled by substantial methodological problems,[2] which make it impossible to positively identify an increase in volume of gold exported from the area, the scholarly consensus is that this era saw increased amounts of the precious metal sold. Since it is not likely to have drawn down previous hordes of the metal, this implies a corollary growth of production.[3] The efforts of nobles and 'big men' to produce more gold in this era were ultimately constrained by the needs of their dependents who dug and panned for the precious metal to provide themselves with food. For so long as these forest farmers centered their subsistence on yam cultivation, their agricultural efforts required much time and toil, which in turn restricted their availability for mineral extraction to only those short, but predictable, periods in the agricultural calendar when the fields required low labor inputs, e.g., weeding.

From the sixteenth century, entrepreneurial-minded men used plantains strategically to expand the range, intensity, and efficiency of their existing gold extraction operations. To gain in labor efficiency, they directed their dependents to pursue food strategies that relied more on less demanding plantains and less on African yams. Leaders and their dependents must have clearly recognized that plantains were much more productive, in terms of labor inputs needed to produce the same volume of bulk starch for food, than were African yams. The organization of labor for agricultural duties was less important, because plantains were perennial plants which didn't require new garden plots to be cleared and prepared each year, as was required for yams. There was no annual 'labor bottleneck' when all male laborers needed to be recruited to clear and prepare the fields for planting. Forest farmers could cut maiden suckers from parent plantain plants that they deemed to have desirable characteristics and replant these suckers on gardens at any time in a year's two rainy seasons. Furthermore, forest farmers could harvest plantains from their plants continuously throughout

[2] Philip Curtin, "Africa in the Wider Monetary World, 1250–1850," in John F. Richards, ed., *Silver and Gold Flows in the Medieval and Early Modern Worlds* (Durham NC: Duke University Press, 1981), 231–68.

[3] Wilks, "The State of the Akan," 360.

the year without interruption and without diminishing the yield. Also, wives and daughters found that plantains were easier to process into food (reduced cooking time meant less dry firewood was necessary); and harvested plantain fruits were less fibrous than the vegetable starch of mature yams and therefore these plantains required less hand-labor to pound into the preferred culinary form of *fufu*.

The strategy of placing plantains at the center of forest foodways thus freed up time that elders and 'big men' could direct toward the extraction of gold. These 'big men' could also locate camps of laborers subsisting on plantain-based diets in immediate proximity to the richest gold deposits in the Gold Coast's most heavily forested and rainfall-inundated landscapes. The advantages of plantains (and especially the highly productive 'French' bunch types that became popular from the late sixteenth century) compared to yams were most exaggerated there.

Forest farmers also benefited from—whether the recognized it or not—an unforeseen boon to their conversion to plantain culture: that their plantain gardens did not hold standing water in which *Anopholes* mosquitoes—vectors the malaria-inducing *Plasmodium* parasite—could breed. Their yam gardens, in contrast, provided ideal breeding grounds in the water caught between the planting mounds that kept the yam tuber dry and free from rot, and the outer ridges of the garden which trapped rainfall needed to nourish yam roots. Perhaps farmers noticed this benefit only in terms of their reduced nightly irritation from pesky biting insects, but they also enjoyed significant unforeseen health benefits. Increasing garden space to plantains therefore directly reduced the population of these annoying bugs and enhanced infant survival rates and the general good health enjoyed by plantain farming populations. These health benefits and their substantial implications for a growing population[4] in the forest suggest that the cultivation and consumption of plantains had critical significance in early Gold Coast demographic—and therefore economic and political—history.

Modern inhabitants of the forest recount in oral traditions that this was an era of new settlement and population dispersal, as their ancestors

[4] Vansina, *Paths in the Rainforests*, 43, 57, 61; James Giblin, "Trypanosomiasis Control in African History: An Evaded Issue," *Journal of African History* 30, 1 (1990), 59–80. Norman Klein has offered a provocative suggestion for the role of malaria in early Gold Coast history and its interplay with human genetics (i.e., sickling traits) and demography; see A. Norman Klein, "Toward a New Understanding of Akan Origins," *Africa* 66, 2 (1996), 248–73.

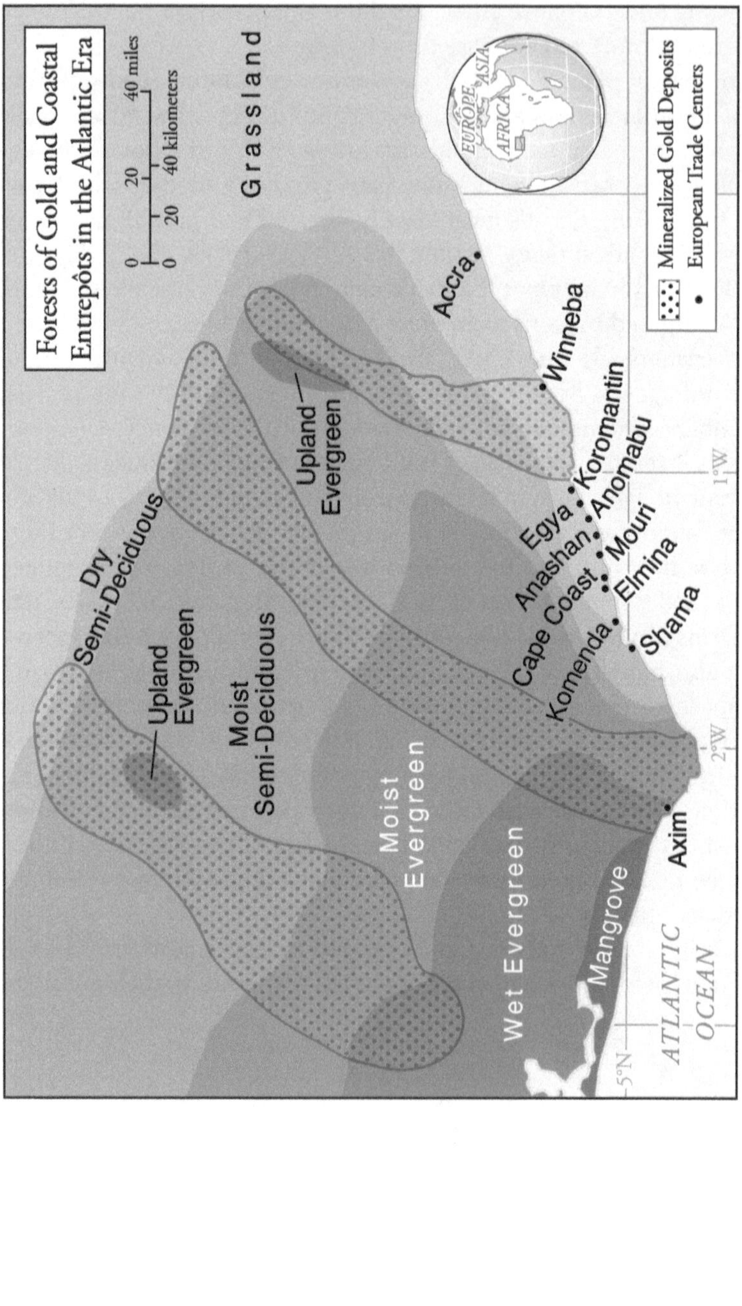


Map 8. Forests of Gold and Coastal Entrepôts in the Atlantic Era (source: mineral data from Kea 1982)

scouted new places to try to make a living.[5] Archaeological research confirms that most of the area's gold mines were opened after 1600—coincident with the adoption of the highly productive French type of plantains.[6] Historians have explained the mechanism behind this expansion, albeit vaguely, as the acquisition of sophisticated gold mining technologies (presumably in the form of persons with expert knowledge) from distant, Sudanic communities with a long tradition of mining.[7] An equally speculative theory would credit Kongo slaves imported to the Gold Coast, among which were some tutored in mineral extraction. These would have been enslaved persons who hailed from the copper mining districts of Mindouli-Boko Songo (just north of the Lower Congo, in the Niari valley) lying in that area which was the most active slaving zone in West Central Africa at the time when slaves from this area were being imported to the Gold Coast. But none of these new strategies could have been implemented had 'big men' not established more efficient, year-round plantain subsistence to support commercial gold extraction enterprises.

Although plantain starch quickly constituted the most important source of energy in the diets of farmers and miners, forest inhabitants never endowed this starchy crop with commensurate cultural value. The social and economic challenge from plantain-growers prompted male elders and entrepreneurial 'big men' to intensify their historic control over yams (see chapter two), which—unlike plantains—required deployment of significant social controls to accumulate the labor necessary to prepare gardens and maximize harvests through social control of the timing of the annual harvest. 'Big men' staved off the potential challenge by promulgating ideas that reified yam culture. They invested substantial resources in celebrating the yam harvest and then transforming that celebration into a demonstration of their own powerful position in the spiritual landscape as pivot between the past and the future; between the living, the dead, and those yet to be born.[8] They also promulgated new ideas—audible now in proverbs—that asserted their new, extensive rights over persons and the fruits of their labor, which they underwrote by extracting an ever increasing share of agricultural and precious metal production.

[5] Wilks, "Land, Labor, Capital."

[6] Anquandah, *Rediscovering Ghana's Past*, 69.

[7] For the slaving, see Miller, *Way of Death*, 56. For Central African copper mining, see Vansina, *Paths in the Rainforests*, 155, 221–24, 236; and Eugenia W. Herbert, *Red Gold of Africa: Copper in Precolonial History and Culture* (Madison: University of Wisconsin Press, 1984).

[8] McCaskie, *State and Society in Pre-Colonial Asante*, esp. chp. 4.

The disjunction between the fruits of farmers' labors and the élites' control over subterranean gold resources was made clear in proverbs which told, 'That underneath the plantain belongs to the king' (Akan, *borɔdeɛ ase yɛ ohene dea*).[9]

These ideas were reflected in the domestic sphere. In the forests and moist grasslands, men endeavored to dominate the cultivation of high-prestige yams, and they disparaged women's crops: water yams (*D. alata*; introduced by the Portuguese from Asia) were said to rob men of their masculinity, and Asian cocoyams were only a stomach-irritating substitute for 'real food.' West African yams remained the most esteemed forest crop in the Atlantic era (and still are today), and forest communities celebrated the end of the West African yam harvest as their major annual celebration. Men claimed sole rights to cultivate this most prized but not productive crop, and attributed phallic imagery to the elongated tubers.

## *In Retrospect*

Locating the introduction of plantains and the rise of their use as a starchy staple in the Atlantic era, rather than before, requires reappraisal of all major historical developments previously described for the early Atlantic era: state formation, population growth, land clearances, and more. Most significantly, the combination argued elsewhere as having occurred in the early Atlantic era[10] of these three major developments into a single 'big bang,' or historical watershed, reliant on the extra labor provided by a small number of imported slaves, in fact depended on: first, the extra labor made available through the relatively diminished labor required to subsist on plantain-based diets; and, secondly, on the clever cultural techniques élite men used to successfully commandeer their dependents' forced labor including offering inducements towards assimilation as full members of the community and using the physical strength of their numbers to force submission to their will.[11] Masses of farmers available for military

[9] The proverb is noted in Kwame Yeboa Daaku, "Trade and Trading Patterns of the Akan in the Seventeenth and Eighteenth Centuries," in Claude Meillassoux, ed., *The Development of Indigenous Trade and Markets in West Africa* (London: Oxford University Press, 1971), 171. I have adjusted Daaku's orthography and translation slightly.

[10] Wilks, "Land, Labor, Capital."

[11] For the application of Gramscian notions of the spectrum of control (from coercion to consent) employed by Gold Coast polities, see McCaskie, *State and Society in Pre-Colonial Asante*.

conscription made for greater difference than the slaves consigned to mining gold.

### *Allada Communities and Culinary Cross-Currents*

In the early Atlantic-era, coastal residents near the early Portuguese trade-forts incorporated Adja- and Fon-speaking strangers into their towns, where the newcomers established themselves as a distinct ethnic community of culinary innovators. These newcomers were slaves imported from the area on the Slave Coast to the east known then to Europeans as Allada, where Portuguese slavers began to procure most of their human cargoes there in the 1550s,[12] as they turned away from previous sources of slaves in Kongo (from the 1510s) and earlier sources at Niger Delta ports (from the 1470s).[13] Although Portuguese factors sold most of these slaves imported from Allada against gold and ivory, they also retained a moderate number of them to serve functions essential to the support of the forts' trading operations, including construction, security, farming, and food preparation. These slaves came to be of critical importance to Portuguese fort administrators towards the end of the sixteenth century, when this remote post received fewer and fewer Portuguese undesirables banished from their home country. Portuguese administrators on the coast supported this slave traffic in contravention of official metropolitan decrees forbidding it. In Lisbon, they feared that slaves sold to West African merchants would strengthen Muslim enemies of Portugal's avowedly Christian empire.[14]

Importation of Allada slaves for domestic chores at these forts continued long after Dutch traders wrested control of them from the Portuguese and became the principal European traders in the 1630s,

[12] Robin Law, *The Kingdom of Allada* (Leiden NL: Research School CNWS, 1997), 87; J. B. Ballong-wen-Mewuda, "Le commerce portugais des esclaves entre la Côte de l'actuel Nigéria et celle du Ghana moderne aux XVe et XVIe siècles," in Serge Daget, ed., *De la traite à l'esclavage: Actes du Colloque International sur la Traite des Noires, Nantes 1985*, 2 vols. (Paris: Société française d'histoire d'outre-mer, 1988), vol. 1, 131–45. For primary source material (and editorial comment) see the remarks of De Marees in Van Dantzig and Jones, trans. and eds., *Pieter de Marees*, 224 with n. 5, 225 with n. 8. See further the Portuguese report of 1620 which described the King of Allada as a 'good friend' of Portuguese traders, and slaves as a significant export item; Castelo Branco in Brásio, ed., *Monumenta*, V, no. 138.

[13] See above, chapter 3.

[14] Van Dantzig, *Forts and Castles in Ghana*, 22, 32n2.

and this practice was copied in turn by English, Danish, and Swedish traders who soon thereafter established trading concerns to compete with the Dutch companies. Their *importation* of foreign slaves to the Gold Coast intensified even as the principal business of these parastatal companies shifted towards *exportation* of great numbers of local captives from this region in the later seventeenth and eighteenth centuries. For example, in 1646 the Dutch *West-Indische Compagnie* (hereafter WIC) purchased 43 slaves at Allada,[15] and again in 1663 the Company sent a vessel with orders to purchase another 100 slaves to be used at the company's factories on the Gold Coast.[16] In 1665–66 the Swedish African Company sent a yacht on a similar slaving voyage to Allada, to acquire servile personnel to man their trading forts on the Gold Coast.[17] Factors of the English East India Company, which administered the English establishment on the Gold Coast from 1657 to 1663, made similar orders to purchase Allada slaves, although these were perhaps unfulfilled.[18] Royal African Company documents from 1678 refer to the presence at Cape Coast of slaves 'not native but brought from Ardra [=Allada] or any other parts.'[19] Allada slaves were at every major English station on the Gold Coast and English factors at smaller satellites hosted teams of slaves on short-term assignments.

European traders preferred imported Allada slaves for domestic chores rather than slaves selected from their inventories of locally acquired trade slaves because, in contrast to the latter, the imported slaves were

---

[15] K. Ratelband, ed., *Vijf Dagregisters van het Kasteel Sao Jorge da Mina (Elmina) aan de Goud Kust (1645–47)* ('s-Gravenhage: Nijhoff, 1953), 245, 287–89.

[16] Koninklijk Institut voor Taal-, Land- en Volkenkunde; Leiden, the Netherlands [hereafter KITLV], manuscript H68, "Resolutie Boeck voor Regeringhe…," entry for 24 August 1663.

[17] Het Nationaal Archief [hereafter NA], Oude West-Indische Compagnie [hereafter OWIC] 47, "Journaal door Louijs Dammaet van een reis me het schip *Prins Willem* naar Guinea," entries for 8 September 1655, 24 February 1656. Although the Swedish company's forts and ships flew that nation's colors, the company was entirely dominated by renegade Dutch personnel and capital eager to trade outside of the WIC monopoly, hence the logical borrowing of the 'Swedish' company from the established practice of Dutch traders in Guinea to procure slaves from Allada for Gold Coast service; for the Swedish company, see Van Dantzig, *Forts and Castles of Ghana*, 23–29.

[18] Law, *Kingdom of Allada*, 91; Margaret Makepeace, ed., *Trade on the Guinea Coast, 1657–1666: The Correspondence of the English East India Company* (Madison: African Studies Program, University of Wisconsin-Madison, 1991), docs 18, 82, 102.

[19] (British) National Archives; Kew, UK [hereafter BNA], T70/77, Minutes of Court of Assistants of RAC, 11 July 1678, "An establishment agreed upon for the better and more orderly government of all of the forts and factories on the Gold Coast or elsewhere in Africa," cited in Law, *Kingdom of Allada*, 91 with n. 522.

socially and linguistically isolated from nearby communities of support and were therefore less likely, or able, to run away.[20] As one English factor at Coromantine in 1661 complained, slaves acquired locally and then tasked with fort duties were 'not serviceable, but allwise running away.'[21] Dutch factors on the Gold Coast reiterated this perception in 1684, complaining that while their locally acquired slaves were 'constantly inclined towards running away and seeking their freedom—in the last month not less that 15 of them have already escaped,' their slaves imported from Allada, 'being strangers, [we]re less likely to run away.'[22] The Allada slaves replaced their lost natal communities with new communities of belonging that were recognized by contemporary European writers. The slaves organized new 'Alata' quarters in the towns that neighbored important European trade forts, including: a small community next to the Danish fort of Christianborg (known locally as Osu), another at Fort Crevecoeur, and a large community next to the English company's fort at James Town (or, Nleshi), all in the area of Accra;[23] a 'slave quarter' next to the Dutch fort at Mouri;[24] a number of 'Adra slaves' at Cape Coast (or, Oguaa [< Fante, *gua*, 'market']) where several European companies had factories,[25] and, presumably, forming a substantial part of those 200 Africans in the town of Elmina who identified themselves in the mid-seventeenth century as 'Portuguese.'[26] These communities saw substantial demographic growth; for example, by the mid-eighteenth century, the company slaves at Elmina had grown to number some 300 persons

---

20 Law, *Kingdom of Allada*, 91. Although such isolation made escape more difficult, it certainly did not make it impossible. In 1681, the English factor at Anomabu reported that one of his Allada men, named 'Quonnu,' had run away to Fetu; Robin Law, ed., *The English in West Africa: The Local Correspondence of the Royal African Company of England, 1681–1699*, 3 vols. (Oxford: Oxford University Press, for the British Academy, 1997–2006), vol. 1, doc. 241. Similarly, in 1687 an English merchant complained that one of his Allada men had run away at Cape Coast; Law, ed., *English in West Africa*, vol. 2, doc. 487. Also, from Charles Fort at Anomabu; Law, ed., *English in West Africa*, vol. 3, doc. 859.

21 Makepeace, ed., *Trade on the Guinea Coast*, doc. 82.

22 Albert van Dantzig, trans. and ed., *The Dutch and the Guinea Coast, 1674–1742: A Collection of Documents from the General State Archive at The Hague* (Accra: Ghana Academy of Arts and Sciences, 1978), doc. 54 ("Resolutions of the Council of Elmina, 7 August 1684") [=NA, OWIC 124].

23 John Parker, *Making the Town: Ga State and Society in Early Colonial Accra* (Portsmouth NH: Heinemann, 2000), 12–14; Reindorf, *History of the Gold Coast and Asante*, 40.

24 NA, OWIC 47, "Journaal door Louijs Dammaet," entries 19 and 20 April 1653.

25 Müller in Jones, trans. and ed., *German Sources for West African History*, 210–11.

26 Olfert Dapper, *Naukeurige beschrijvinge der Afrikaensche gewesten…* (Amsterdam, 1668), 446, 447; Hair, Jones, and Law, trans. and eds., *Barbot on Guinea*, vol. 2, 381.

and constituted half of the Dutch company's Gold Coast slave personnel.[27] Castle slaves earned wages, owned property, were not without rights and opportunities to redress mistreatment, and they lived, worked, socialized, and sometimes entered into family relations with Europeans and indigenous Gold Coast people. Fante-speakers generated their own names for these outsiders, which indicated their perception of them as resident aliens: as *brofunkoa* (Fante, 'white men's slaves' [<*borɔfo*, white men + *ŋkoa*, 'slaves']) and *brofomba* (Fante, 'white men's children' [<*borɔfo*, white men + *mbá*, 'children']).[28]

These Allada communities were uniquely positioned to become cultural intermediaries with the Atlantic and culinary innovators. They acquainted themselves, through their servile role, with the altogether foreign techniques, imported ingredients, and culinary forms associated with a European mess hall. From the moment of their arrival on the Gold Coast, Allada slaves also catered to their own culinary preferences. Residence of many outside the forts gave them opportunity to try to satisfy their own cravings for home cooking, which they did by reinterpreting their native cuisine within the constraints of locally available produce.

### *Baked Bread and Biscuits*

Enslaved castle cooks baked approximations of European styles of fresh bread for fort personnel to consume in a fresh state, and they prepared dry bread foods for storage and consumption as shipboard provisions. Presumably, their knowledge about European breads was derived directly from Portuguese women who had been banished to West African service as *degradadas* from the outset of the Portuguese imperial scheme in the late fifteenth century [see above, chp. 3]; records from the Portuguese forts of the Gold Coast record several banished women in service as 'kneaders' for bread.[29] Slaves from Allada who assumed the jobs of castle cooks in the later sixteenth century and, subsequently, those people living in communities immediately next to the trade forts borrowed the recipes for making these breads and adapted the recipes to local ingredients, and

[27] NA, Archief van de Nederlandsche Bezittingen ter Kuste van Guinea [hereafter NBKG] 287, 16 January 1758; NA, NBKG 317, 13 March 1784.

[28] Ansu Data and R. Porter, "The Asafo System in Historical Perspective," *Journal of African History* 12, 2 (1971), 279–97; Arthur Ffoulkes, "The Company System in Cape Coast Castle," *Journal of the African Society* 7 (1907–8), 262–77.

[29] Hair, trans. and ed., *East of Mina*, 36 with 91n195–96.

names for breads cooked in European styles entered the coastal trading vocabulary as loan words.

Castle cooks prepared fresh bread in the same general manner as done by European bakers. They mixed flour with a greater measure of liquid and a inoculate of active yeast (usually to be found in the form of a piece of old bread or the frothy leftovers from beer-brewing); sometimes they also added a bit of animal fat (butter, milk, eggs) or vegetable oil.[30] They kneaded this mixture until it was soft and stretchy, then left it to rise. Thereafter the risen dough was divided into desired portions and placed into a hot oven. On the Gold Coast, castle bakers were restricted to making rudimentary versions of European breads because of the different ingredients and equipment available to them in West Africa. Local versions adulterated, or completely replaced, imported European grain (wheat, rye, barley) flours with the meal of locally grown produce—principally maize and millet—and yeast was as often procured from sugar-spiked palm wine as from old bread,[31] and, finally, if a fatty additive was desired, they replaced butter, milk, and olive oil with palm oil. Europeans stationed at Gold Coast forts counted on these loaves for their 'daily bread.'

Along the entire coast, speakers of local languages borrowed directly from the Portuguese word *bolo*, ('loaf') as their generic name for this oven-baked bread, and contemporary European visitors to the coast noted versions of this word in their lists of vocabulary needed to conduct trade there. Locals' use of this term was probably further reinforced through subsequent contact with Dutch, Danish, and German traders on the coast, who referred to such a bread in their own native tongues as the similar sounding *bol* ('bread roll'). Gã-speakers also borrowed a name for such bread from the Dutch and Danish traders in their midst, who named this *brood* or *brød*, respectively; Gã-speakers reinterpreted these names into the sound of their own speech (again showing the typical /r/ > /l/ sound change) as *blodo*, a synonym for their Portuguese-derived term [see Word List 5].

Locals made further distinctions between those breads baked from unadulterated wheat flour and the local, primarily maize-based variants.

30 Such 'shortening' (to use bakers' parlance) makes the loaf more tender, but the chemical explanation of how added fat accomplishes this is yet unknown; McGee, *On Food and Cooking*, 303–4.

31 On the Gold Coast cooks' use of old bread as their yeast starter, see Brodie Cruickshank, *Eighteen years on the Gold Coast of Africa, including an account of the native tribes*... 2 vols. (London: Hurst and Blacket, 1853), 275; on the use of sugar-enriched palm wine, see Field, "Gold Coast Food," 114.

Word List 5. Comparison of Vernacular Names for Loaf Breads

| Language | Dialect | Loaf Bread | Comment |
|---|---|---|---|
| Akan | Fante | paanoo | |
| | Nzema | | |
| | Akuapem | abodoo | < *bolo*, Portuguese, 'loaf' |
| | | bodobodoo | reduplication |
| | Asante | brodo | |
| | Brong | | |
| Guang | Coastal | | |
| | Hill | | |
| Gã-Dangme | Gã | blodo | < *brood*, Dutch, 'bread' and/or Danish *brød* |
| | | aboloo | < *bolo*, Portuguese, 'loaf' |
| | | bodobodo | reduplication |
| | Dangme | | |
| Gbe | Ewe | abolo | < *bolo*, Portuguese, 'loaf' |
| | | yevubolo | < lit, 'Europeans' *bolo*'; term specifically refers to wheat bread |

The preponderance of maize-flour breads in the Atlantic era supported the locals' attribution of generic meaning to *bolo*-derived terms. They also borrowed from the Portuguese word *pão* (generic, 'bread'), which they reinterpreted in the phonology of their own speech (Akan, *pããnõõ*; Gã, *akpɔnɔ*; Ewe, *akpɔnɔ*) as yet another synonym for bread made primarily from locally grown grain. To identify the wheat breads of élite consumption they used reduplications (e.g., *bodobodo*) that indicated intensification, or they created neologisms which identified wheat breads as Europeans' version of local food (e.g., Ewe, *yevubolo*, lit., Europeans' *bolo*).

Castle cooks also prepared approximations of 'hard tack' or 'ships' biscuit' to provision ships departing from the coast. These were dense, dry, unleavened breads particularly well suited to transportation and storage. The density of these breads made them less likely to be penetrated by insects once exposed to moisture—either through poor storage or which was inevitable once their storage containers were opened. This density also allowed loadmasters to reduce the space needed to store provisions and instead equip ships with more trade goods and matèriel. Such breads were the maritime interpretation of Europeans' hard, dry breads eaten by soldiers garrisoned on land; there, leavened fresh bread was cooked twice to completely rid it of moisture on which moulds and insects could grow and induce quick spoilage. In France, quartermasters prized these biscuits (< French, *pain biscuit*, lit., 'twice-cooked bread'; cf. the loan-translation in German, *zweiback*): bread was baked, sliced, and then the slices were put back into the oven until they were completely dry.[32] Moisture extraction not only prevented spoilage, but also made the breads substantially lighter and therefore cheaper to transport overland.

At the European trade forts of the Gold Coast, cooks prepared ships' biscuits for the consumption by crews on outbound voyages. While Portuguese traders controlled the fort at Elmina before 1637, bakers made these breads by using exclusively maize flour; these biscuits sustained sailors transiting between Gold Coast and São Tomé and Angola and were credited with having a shelf life of 'three to four months.'[33] The bakers at the Portuguese's rivals' forts at Cape Coast used only millet flour.[34] Along the entire coast, bakers and their customers referred to these locally made approximations of ships' biscuits by a Portuguese derived term; a mid-seventeenth-century Dutch guide to trading on the Gold Coast advised that the name *escotte* (< Portuguese, *biscoito*) was the local name used in commerce.[35] By the end of the eighteenth century, this food had fallen into near obsolescence, and the term was largely forgotten; instead locals reincorporated all such biscuits within the semantic of local bread and thus identified them with variations of the generic term *pããnõõ*.[36] Only during the

---

[32] McGee, *On Food and Cooking*, 320. This is a very much different food from modern Anglophone interpretations of biscuits: in Britain as a (often) sweet-tasting, dry-texture, single-baked bread product (typically taken with, and sometime even dunked in, tea); and in the United States as small, moist bread which is charged with much baking powder to give it a fluffy, soft rise, and is meant to be eaten fresh, and usually split in half to hold sliced meat or so as to spread with butter and/or preserved fruits. It also different from the modern Italian *biscotti*, which is a sweet (and often almond-studded) twice-baked dessert bread. The closest modern approximation of the original recipe is perhaps the bland and undecorated bread known in Dutch as *beschuit* (clearly a French loan word adopted—as part of an entire semantic field of military and bureaucratic matters—by Dutch speakers during the period of Napoleonic overrule). This same dried bread is known, though somewhat obscurely, in modern English as 'rusk' (< Spanish and Portuguese *rosca*, 'a coil; a twisted roll of bread').

[33] Van Dantzig and Jones, trans. and eds., *Pieter de Marees*, 113, 40; see also Hemmersam's remarks in Jones, trans. and ed., *German Sources for West African History*, 111.

[34] Jones, trans. and ed., *German Sources for West African History*, 209.

[35] Jones, trans. and ed., *West Africa in the Mid-Seventeenth Century*, 221.

[36] George A. Robertson, *Notes on Africa, particularly those parts which are situated between Cape Verde and the River Congo*... (London: Sherwood, Neely and Jones, 1819), 281; Capt. John Adams, *Sketches taken during ten voyages to Africa between the years 1786 and 1800*... (London: Frank Cass, 1823), 41.

heyday of British colonial overrule in the Gold Coast in the mid-twentieth century did biscuits (of the English, single-baked confectionary sort) again attract limited attention of locals, who borrowed anew from European languages when they named the food *bisket* [see Word List 6].

Few indigenous cooks made baking these sorts of Atlantic-era breads part of their culinary repertoire. Therefore, few who ate from their hearth ever made baked breads the daily starchy staple of their diets, and they never affixed to the loaves the cultural importance Europeans did. Those Gold Coast cooks who did regularly bake were those who lived lives interwoven with resident Europeans—as slaves, or the informal 'wives' and children kept by European men stationed at the forts. Perhaps some among the dependant communities living in the shadows of the fort baked bread because they recognized that there was a limited but established market of consumers desiring such foreign food, but not for themselves. Most locals always considered baked bread to be an impractical foreign food, and associated its consumption with foreigners. In the 1610s, scarcely any

Word List 6. Comparison of Vernacular Names for Biscuits

| Language | Dialect | Biscuit | Comment |
|---|---|---|---|
| Akan | Fante | c. 1650 *escotte* (Annon. MS) | < Port. *biscoito*, 'biscuit' |
| | | c. 1818 *panoo* (Robertson) | < Port. *pão*, 'bread' |
| | | c. 1823 *panoo* (Adams) | |
| | | bisket | < English, 'biscuit' |
| | Nzema | | |
| | Akuapem | paanoo | < Port. *pão*, 'bread' |
| | Asante | | |
| | Brong | | |
| Guang | Coastal | | |
| | Hill | | |
| Gã-Dangme | Gã | akpaanɔɔ<br>(alt.) akplaanɔɔ | < Port. *pão*, 'bread' |
| | Dangme | | |
| Gbe | Ewe | akpaanɔ | < Port. *pão*, 'bread' |

indigenous cooks worked ovens for baking bread,[37] and as late as the 1780s, ovens and baking were still limited to those coastal communities immediately next to the European forts.[38] Elsewhere, family cooks must have regarded baking bread as a terrible waste of their energies. Not only did baking require a substantial, initial investment of labor to construct an earthen oven, but all baking thereafter required a moderate, recurring (daily) effort to make and tend to a second—and very hot—fire in addition to the main, low-heat fire over which they cooked all soups and stews and other foods. To bake, cooks would pre-heat an oven by burning a load of wood inside of the earthen enclosure; once the fire had reached its peak temperature, cooks then swept out the embers and ashes, inserted their loaves, and tried to minimize heat escape via the opening by covering this vent with a wet cloth. The loaves cooked while the oven cooled.[39] From the time the loaves were ready, bakers had only a very narrow window of time to serve or sell their loaves before the ever-humid tropical air rendered them stale and inedible.

Those lacking the wherewithal to eat baked bread relied on easier and cheaper means to turn local grains into a starchy meal. They mixed grain flour with water and placed their dough at the margins of the fire, where the bread cooked on heated rocks or ashes.[40] Such simple breads—common worldwide, and known in English as ashcakes—were the mainstay of the peripatetic, as their preparation required no specific cooking implements to be carried and no significant preparation time (e.g., for leavening) once the immediate destination was reached.

### *Kenkey*

As coastal consumers turned increasingly to maize as their dominant starchy staple, cooks began to experiment with applying a number of cooking techniques that introduced culinary variety to their otherwise monotonous maize-based diets. Initially, it was Allada slaves residing in

---

37 Brun noted that, 'they have no ovens'; in Jones, trans. and ed., *German Sources for West African History*, 85.

38 Isert remarked, 'Baking is something learned from the Europeans, since one never finds a baking oven in the places where no European has stayed . . .'; Selena Winsnes, trans. and ed., *Paul Erdmann Isert's Letters on West Africa and the Slave Trade* (Oxford: Oxford University Press, for the British Academy, 1992), 125.

39 Van Dantzig and Jones, trans. and eds., *Pieter de Marees*, 112n7; Field, "Gold Coast Food," 14.

40 Van Dantzig and Jones, trans. and eds., *Pieter de Marees*, 112.

the Gold Coast trading towns who demonstrated particular expertise in the preparation of new foods made from maize, although they had no prior experience whatsoever with the American grain; as late as the 1680s, maize was just being introduced to the Slave Coast.[41] But Allada slaves transported to the Gold Coast drew upon the rich repertoire of recipes developed for millet and sorghum grain, around which their natal communities had centered their foodways since the distant past, to prepare new foods from maize.

Slaves brought from Allada probably spoke of their bread as something akin to *kaná* or *kalã*, which are the current names for similar, indigenous 'breads' in the area of Porto Novo (Benin).[42] These terms coalesced in the coastal traders' *lingua franca* in a modified form ancestral to the modern 'kenkey,' the maize 'bread' which has been the staple food of the Gold Coast littoral since the seventeenth century. The Gold Coast version of this name was first recorded c. 1600 in Flemish orthography as *kangue*,[43] which was probably a somewhat faithful reinterpretation of the name borrowed by those involved in Atlantic trade.[44]

The epicenter of this culinary revolution was Elmina Castle, where Allada women who worked as fort bakers also produced fermented maize breads for their own consumption and for sale at neighboring markets as an exotic foodstuff. In the early seventeenth century, the Allada cooks of Elmina were hauling their products at least as far as the street market at

---

41 Hair, Law, and Jones, trans. and eds., *Barbot on Guinea*, 645n1.

42 These local breads are now made from maize and are steamed like kenkey.

43 Van Dantzig and Jones, trans. and eds., *Pieter de Marees*, 224 with note 5; 225 with note 8. This /ng/ sequence probably represented a sound—perhaps a Flemish approximation of a nasalized semi-vowel /n/—rather a sequence of consonants, as this sequence is most rare in languages spoken in the area which was the Slave Coast. Further, Flemish-speakers characteristically make a 'soft' /g/ sound when compared to their linguistic brethren to the north, who use a hard, guttural sound. But perhaps De Marees's Flemish orthography was changed when his manuscript was set into type in the northern Netherlands; for a contemporary example of such orthographical change, see La Fleur, trans. and ed., *Pieter van den Broecke's Journal*, 15.

44 The etymology of the name kenkey (and its Atlantic-era renditions) has typically confused historians who looked for indigenous origins. Typically they have identified the modern Akan *kãŋkyew* ('cake') as the progenitor, but this explanation suffers from documentary and linguistic weakness. Early documents make it clear that kenkey was the name used by Europeans for the local 'bread,' while indigenous people used different terms; see for example Isert's remarks in Winsnes, trans. and ed., *Isert*, 125. And linguistically, this sound change /ky/ > /k/ is not attested in Akan phonology (i.e., velar plosive to alveolar fricative). The IPA representation for the Akan /ky/ is tɕ, akin the initial sound in the English word "change."

Cape Coast fifteen kilometers to the east.[45] The exotic origin of kenkey is further evidenced by documents from English factors at trade-forts on the Gold Coast who specifically recruited (female) Allada slaves to have them make *kenkey* in the 1680s for slaves held at the lodge at Anashan (nowadays Biriwa) and Egya.[46]

Cooks soaked dried maize grains in a pot of water for a day, until they were soft. They then decanted the water and placed the swollen maize, handful by handful, on a grinding stone. Once ground, the mash was mixed again with water and left to ferment spontaneously for several days. Cooks then divided this fermented dough in half; while one part rested, they mixed the other with water and salt and cooked to make the firm mash known as *aflata*. The two parts were then recombined, molded into fist-sized portions, and wrapped in waxy leaves. Cooks then exposed these parcels to the dry heat in fort ovens or near cooking fires until they were sufficiently firm enough to handle and eat by hand and dip into soups and sauces.

Akan- and Gã-speakers living in the coastal fort communities adapted the foreigners' recipes for this sort of 'bread' and designated their domesticated versions with neologisms that reveal their initial impressions of this food. Gold Coast cooks abandoned dry heat preparation and instead used the application of moist heat (i.e., in a pot of steaming or boiling water, which spared them the effort of making a new fire). Gã-speakers invented a new name, *kɔmi*, (from *kɔ*, 'bite' [habitual] + *mi*, 'swallow' [habitual]) which marked their appreciation of this bread as a fast food. Gã-speaking women prepared their breads by wrapping the fermented maize dough together with a sprinkle of sea salt inside a maize husk. Fante-speaking market women prepared their dough in similar fashion, but preferred to bundle their fermented dough without the savory seasoning inside a green *Musa* (i.e., plantain or banana) leaf, which imparted a desired perfume to the contents of the parcel. The Fante-speaking consumers named this food *dɔkunu*, probably capturing their favorable impression of the food as pleasing them (Akan, *dɔ kɔŋ*, '[it] pleases')[47] This term and culinary form were borrowed by speakers of other Akan dialects (e.g., Asante, Akuapem), including those transported to plantations of the Caribbean, where the food and the name for it continue to pass over the lips of their modern

45 Van Dantzig and Jones, trans. and eds., *Pieter de Marees*, 62.

46 Law, ed., *English in West Africa*, vol. 1, docs 119, 378.

47 The phrasing of the experiential verb reverses the normal word order (c.f., *ne kɔŋ dɔ ade*, 'he desires/wants food'). See also the Akan adjective, *dɔkɔɔ*, 'palatable.'

descendants.[48] Nzema-speakers near the trade forts of the far west also invented a new name, *fomfom* (perhaps a reduplication of *fɔŋ*, alt. *fãm*, 'to become emaciated'),[49] indicating their initial recognition of this bread as provision for company slaves and those underfed and malnourished captives destined for export [see Word List 7].

Cooks as well as the their family consumers and retail clients may well have noticed that fermented foods had health benefits as well as engaging flavor. They might have noticed that fermented maize foods reduced the incidence with which they felt ill and weakened. Maize foods that had been processed by lactic acid fermentation created biochemical conditions inhospitable to pathogens and mildly increased the availability of nutritious elements in them.[50]

These breads also resisted spoilage for several days even in tropical weather, and this extended stability reduced otherwise wasted amounts of retail inventories, basic ingredients and cooking labors.[51] Such attributes favored the universal embrace of kenkey as an important starchy food among all coastal communities. From the later seventeenth century, kenkey was so central to their foodways that *per diem* disbursements for the subsistence of laborers, craftsmen, soldiers, and canoe-men were known as 'canky money.'

---

[48] This food is known in Jamaica as *dokunu* as well as *boyo* (< early Gold Coast *bolo*) and 'blue drawers' (a neologism which takes note of pale the blue-green hue the leaf wrapper leaches into its contents and the careful way the parcel is folded like boxer shorts, i.e., underwear); and in Barbados as *dokunu* and as *conkie* (< kenkey). On the arid island of Anguilla, enslaved persons fashioned *conkies* by soaking and grinding (imported) maize and wrapping the mash in leaves of the sea grape (*Coccoloba uvifera*). The Aluku (Boni), Saramaka, and Ndjuka maroon communities of French Guiana and Suriname know wrapped maize dumplings as *doku*; see Richard Price, "Subsistence on the Plantation Periphery: Crops, Cooking, and Labour among Eighteenth-Century Suriname Maroons," in Ira Berlin and Philip D. Morgan, eds., *The Slave's Economy: Independent Production by Slaves in the Americas* [special issue of *Slavery and Abolition* 12 (1991)] (London: Frank Cass, 1991), 118, 126n30. The Jamaican synonym *boyo* for this food is also heard in the Palenquero maroon community of Colombia as *bollo limpia*.

[49] An alternative etymology might be a reduplication of the Akan verb, *fãm*, 'to tie up.' Cf. Akan, *ɔfam*, alt. *tetare*, 'a flat cake,' which is now made from ripe plantains mixed with flour—usually made from rice—and some onions and then fried; this is altogether a different food from kenkey. See Fields, *Gold Coast Food*, 16.

[50] P. Mensah, A. M. Tomkins, B. S. Drasar, and T. J. Harrison, "Antimicrobial Effect of Fermented Ghanaian Maize Dough," *Journal of Applied Bacteriology* 70 (1991), 203–10.

[51] Pius Fru Nche, "Innovations in the Production of Kenkey, a Traditionally Fermented Maize Product of Ghana: Nutritional, Physical and Safety Aspects" (PhD thesis, Landbouwuniversiteit Wageningen, 1995).

Word List 7. Comparison of Vernacular Names for Kenkey

| Language | Dialect | Steamed Maize Food | Comment |
|---|---|---|---|
| Akan | Fante | c. 1600 *kangue* or *kanquies* (De Marees) | < *kaná* or *kalã*, Adja and Fon, 'bread' |
| | | c. 1624 (Brun) *kancty* | |
| | | c. 1662 (Müller) *cántje* | |
| | | c. 1721 (Atkins) *canky* | |
| | | c. 1732 (Phillips) *cancy* | |
| | | c. 1758 (Thompson) *cankee* | |
| | | c. 1780 (Isert) *kankis* | identified as the European name for this food |
| | | (modern) kenkey | synonym widespread in coastal languages of West Africa |
| | | c. 1818 (Robertson) *duconnöo* | < Akan, dɔ kɔŋ, '[it] pleases' |
| | | dɔkunu | |
| | Nzema | fomfom | reduplication of fɔŋ, alt. *fãm*, 'to become emaciated' |
| | Asante<br>Brong | dɔkunu | < Fante |
| Guang | Coastal<br>Hill | | |
| Gã-Dangme | Gã | c. 1780 *kummy* (Isert) | kɔ, 'bite' [habitual] + *mi*, 'swallow' [habitual]) |
| | | kɔmi | |
| Gbe | Ewe | dɔkunu | < Akan |
| | | kɔŋ | < Gã |
| | | tim | |

### *Opportunities Brewing*[52]

Allada cooks extended their expertise with grain fermentation to make beverages, some of which boasted alcoholic content. Perhaps the Allada of the Gold Coast first began to brew merely in order to quench their own thirst for a drink that reminded them of home. But brewers also then met the substantial retail potential presented by maritime traders thirsty for any tipple as well as Gold Coast consumers eager to experiment with new, Atlantic cuisine.

Allada slaves drew on their recipes and techniques of their homelands to make beer[53] at the European forts on the Gold Coast. Brewers immersed previously dried grains in water for a few days until they were fully softened, then removed them and placed the soaked grains in wrappers of plantain leaves until the seeds began to sprout. This germination triggered a natural response in the seeds to release enzymes to break down their complex starches into simple sugar molecules. Brewers then arrested this germination process by drying the sprouted seeds and then grinding them to make a sugar-rich meal, or 'malt.' Brewers then tipped the malt into a pot of boiling water to dissolve the sugars and left this mixture to cool; then they poured it through a finely woven basket that trapped insoluble remains. Brewers then left the filtrate, or 'wort,' to stand overnight and assume the slightly sour taste (thanks to airborne introduction of certain organisms, including *Geotrichum candidum* and *Lactobacillus* spp.) that drinkers desired. The following day, brewers again brought the pot to boil in order to further concentrate its sugars and deepen its flavor,[54] and then again left it to resume ambient temperature, whereupon they inoculated the concentrate with a yeast (*Candida* spp.) starter taken from a previous batch of brew. These yeasts converted the concentrated sugars overnight into alcohol by the well-known process of fermentation.[55] From the next

---

[52] I thank Wendell Iverson, formerly of Heineken Brouwerijen (Zoeterwoude NL), for sharing his expertise in brewing, African products, and beer tasting with me.

[53] These West African brews were technically 'ales' rather than 'beers,' as they lacked the hops that differentiate the former from the latter. The term beer is used here because of its current use in conventional English as a generic term, and because the terms were used interchangeably in Atlantic-era descriptions of the African beverages.

[54] The second heating also had the beneficial effect of killing illness-inducing spores that had been awakened from their dormancy by the heat and moisture of the first cooking.

[55] J. A. Ekundayo, "The Production of Pito, A Nigerian Fermented Beverage," *Journal of Food Technology* 4 (1969), 217–25. Müller offered the most lucid description of mid-seventeenth century brewing techniques of the Gold Coast: 'They tip a pile of *millie* into

morning, the beer was ready to be consumed, which needed to be done quickly before the batch spoiled.

Allada slaves brewed with whatever grain was available in the vicinity of the trade-forts of the Gold Coast. But they exhibited a consistent preference to use millet and sorghum, which were the standard ingredients remembered from old, Slave Coast recipes. Although Allada brewers of the Gold Coast experimented with substituting maize, the most abundant domestic resource, for those grains, Allada communities preferred to invest additional labor to make and keep special gardens of millet and sorghum reserved for their beer making. Perhaps they were motivated by sentimentality for the ancestral grain or perhaps it was the superior taste millet malt produced. Maize malt is universally considered to make poor tasting beers and is generally avoided by those brewers who have different grains available. By the late seventeenth century, millet cultivation and culture on the central coastlands was closely identified with brewing rather than producing starchy foods.[56] European visitors identified Allada communities as the brewers of the Gold Coast.[57]

If native corollaries ever indeed existed, the new, imported brewing tradition appears to have overwhelmed indigenous techniques and identifying terms among coastal communities. There is no evidence of earlier beer brewing on the central coast. Archaeological perspectives are unable to distinguish between high-lipped pots that residents used both to hold water and to brew. Oral traditions are silent about the antiquity of beer, perhaps because of its relatively recent transplantation to the Gold Coast, or because since colonial times the intrusion of international beer companies has squashed household brewing, or because those élites who sponsored the oral arts wished to celebrate the (putative) discovery of palm wine—the powerful medium which they used to access the power of the ancestors—and remain silent about the new alcoholic drink to

---

a large dish full of water and leave it lying there till it sprouts and germinates. They then dry it in the heat of the sun. Then, when they want to boil beer, they place large pots of water over the fire; and when it becomes boiling hot, they tip the malt into it and let it cook for several hours. When they think it has cooked enough they let it run through sieve-baskets into pots underneath, and again place it on the fire. This they do repeatedly until it acquires a brown color and the taste of beer, whereupon they leave it to stand in the posts till it becomes cold'; in Jones, trans. and ed., *German Sources for West African History*, 210. See also De Marees's brief description in Van Dantzig and Jones, trans. and eds., *Pieter de Marees*, 113.

56 Jones, trans. and ed., *German Sources for West African History*, 111, 207.

57 See, for example, Hemmersam's comments in Jones, trans. and ed., *German Sources for West African History*, 111; and Bosman, *Nauwkeurige Beschryvinghe*, vol. 2, 248.

which they could not limit access. Finally, linguistic analysis is not helpful in revealing a substantial indigenous brewing history.

Coastal brewers produced a final product that resident Europeans compared favorably with the low-alcohol beers (known in their contemporary description as 'small' or 'half' brews) of their native countries. Brewers could also skip the final fermentation stage and produce non-alcoholic beverage, as was noted by some European visitors.[58] These 'malt' drinks are the direct ancestors of those now made by international conglomerates and marketed in this area as a healthy choice for growing children. The purported healthfulness of the malt drink may have indigenous historical antecedents, as these drinks contain sugars and minerals akin to those used to rehydrate children sick with diarrheal diseases, the leading cause of child mortality in Africa.

Outside the fort, local women copied these exotic brewing techniques (or used them to enhance their own existing recipes) and they borrowed the word the new brewers used to describe their product (Adja and Fon, *ahàn*, 'alcoholic drink')[59] and transformed the sound of this term only slightly into their own speech as *ahei.*[60] Autochthons familiar with the prominent local form of alcoholic drink, palm wine, and the name for which (Akan, *nsa*) was synonymous with alcohol, applied this new term solely to designate such the new, grain-based, fermented drink [see Word List 8].

European visitors came to know this drink brewed from grains grown in Atlantic Africa as *pito*. They adopted this name into their own maritime vocabularies by borrowing from the first African brewers with whom they came into contact, in Senegambia, and who shared a cognate of this word for beer with grain farmers throughout the Sudan (cf. Hausa, *fitoo*). Europeans then used the term, *pito* (spelled variously in early European documentation as *pitou, pitouw*, and other variations, and sometimes likened to wine), in their own way to describe to those local brews they subsequently encountered elsewhere in coastal West Africa,

---

[58] See Rømer's comment on non-alcoholic brew in Winsnes, trans. and ed., *A Reliable Account of the Coast of Guinea*, 235. The notation elsewhere (127n30) that all local brews were not fermented and therefore non-alcoholic is, to my mind, mistaken.

[59] Cf. compound words in Adja and Fon which identify the particular grain fermented to make certain alcoholic drinks: *li'hàn*, 'millet beer'; *abo'hàn*, 'sorghum beer.'

[60] Recorded as *akay* in circa 1663 by Hemmersam in Jones, trans. and ed., *German Sources for West African History*, 111; and as *ahei* by Müller in Jones, trans. and ed., *German Sources for West African History*, 211, 353.

Word List 8. Comparison of Vernacular Names for Brewed Beverages

| Language | Dialect | Local Brew | Comment |
|---|---|---|---|
| Akan | Fante | c. 1600 *pitouw* (De Marees) | < African maritime trade vocabulary |
| | | c. 1663 *akay* (Hemmersam) | < Adja or Fon *ahàn*, 'alcoholic drink' |
| | | c. 1666 *ahei* (Müller) | |
| | | ahai | |
| | Nzema | | |
| | Akuapem | ahai | |
| | Asante | ahai | |
| | Brong | | |
| Guang | Coastal | | |
| | Hill | | |
| Gã-Dangme | Gã | c. 1780 *mada* (Isert) | ŋmadaa, 'millet beer'; from Gã, from *ŋma*, 'millet' + *daã*, 'drink' |
| | | ŋmadaa | |
| | Dangme | | |
| Gbe | Ewe | ahai | |

regardless of whatever names indigenous people used to identify their grain-based drink.[61]

Some Europeans approached local beers with trepidation. Drinkers complained that consuming local beer gave them 'wind' (i.e., gassy, bloated, and flatulent),[62] probably as a result of sorghum malts. These left complex sugars that the introduced yeast could not consume to remain in the brew, and these molecules passed undigested though the stomach of the drinker and into the bowel, where resident bacteria fed on them and produced gasses as a by-product of their digestion.[63] Europeans also fretted about the local beer's cloudiness, which they considered to be an

[61] Van den Broecke described it as *'wijn de po'*; see La Fleur, trans. and ed., *Pieter van Broecke's Journal*, 38 with n. 4.

[62] Jones, trans. and ed., *German Sources for West African History*, 210; Winsnes, trans. and ed., *A Reliable Account of the Coast of Guinea*, 235.

[63] Similar sugars in beans have the same effect. These sugars in sorghum beer are different than the cause of modern lager (barley malt) drinkers' lament, blame for which lies on the (over-)carbonization of their brew.

indication of the drink's unhealthy nature, but which was in fact, a result of the high iron content of African surface water and which in fact provided beneficial nutrition to drinkers. Furthermore, Gold Coast beer drinkers and European imbibers recognized that drinking fermented brews was eminently safer than suspect local water supplies that were often home to waterborne pathogens capable of inducing severe, even deadly, gastro-intestinal maladies.[64] But, given the scarcity of European brews that reached the Gold Coast unspoiled, most were not shy about indulging in the local product. Those few indigenous beer drinkers who sometimes had access to imported European beers[65] considered the foreign drink to be significantly distinct enough from their own that they invented a name for it which indicated its connection with foreigners (Akan, *borɔfo ahai* or *borɔfo nsa*, 'white folks' *ahei* or *nsa*'; Gã, *blofo ŋmadaa*, 'white folks' *ŋmadaa*').

In the environs of Accra, Gã-speaking brewers retained their own brewing traditions, focusing the production of a millet-based alcoholic drink they knew as *ŋmadaa* (from Gã, *ŋma* 'millet' + *dãa*, '(generic) drink'). The dry environment of the Accra plains and the resultant scarcity of moisture-loving palm trees elsewhere tapped for fermentable juice on the Gold Coast had probably stimulated indigenous brewing technologies in the deep past. Even as maize became the dominant grain of the area and sometimes used in beer making as well as most other foods, elders reserved a special role for millet beer in their rituals.[66] The appearance of foreign brewing traditions and growing imports of strong alcoholic drink—rum from Brazil, *jenever* (Dutch 'gin,' whence came the British corruption in language and recipe for 'London dry gin') from the Netherlands, and assorted European brandywines—gave south-east communities unprecedented amounts to drink and the spiritual access they thought alcohol made possible.

Finally, by brewing and marketing beer, women of the Gold Coast created space for independent income accumulation outside existing dominance of male elders over alcohol production from palm trees that grew in the groves they controlled. The establishment of brewing in the towns

---

64 Bosman, *Nauwkeurige Beschryvinghe*, vol. 2, 248. Cf. Barbot's contrary assertion in Hair, Law, and Jones, trans. and eds., *Barbot on Guinea*, 634, 645n2.

65 See, for example, Rømer's observations in Winsnes, trans. and ed., *A Reliable Account of the Coast of Guinea*, 158.

66 Akyeampong, *Drink, Power and Cultural Change*, 33.

of the central coast gave non-elders—junior men and women, slaves and free alike—access to alcohol and the spiritual power they understood it to provide. They prized this access, and certainly the physical pleasure of imbibing, enough to continue to brew beer from maize and sorghum in the Americas, where they tutored European women in brewing non-European starches.[67]

### *Sowing and Savoring Wealth*

Kings and 'big men' of the coastal grasslands and, later, independent entrepreneurs made themselves rich in Atlantic era by sowing maize seeds to cater to European traders' need for provisions. From the era of earliest Portuguese visitation of the Gold Coast in the fifteenth century, chiefs of the major coastal ports (and the kings of the states with capitals in the hinterland that sometimes exercised authority over the littoral) supplied visitors with local foodstuffs in order to sustain them during their stay and to provision their onward voyages. African notables used these contacts with foreigners to enhance their own wealth by trading domestic provisions for exotic goods they could distribute to others, thereby creating personal networks of the clients who signified wealth in the domestic economy.[68]

The establishment in ensuing centuries of permanent settlements of other European traders in growing numbers created demands for foodstuffs that challenged the ability of local notables to satisfy simply by diverting small surpluses reaped locally in years blessed by favorable environmental conditions. Coastal nobles therefore developed means of extracting additional labor from their subjects; they extended the traditional tithe (Akan, *abusa*) of part of marine catches[69] (which had been formerly the principal saleable product of their dominions) to claim shares of their subject's

---

[67] Sarah Hand Meecham, "Going Over and Coming Back: Gender, Technology, and Household Cidering in England and the Chesapeake 1690 to 1760" (unpublished conference paper; Omohundro Institute of Early American History and Culture Conference, 14–16 June 2002, College Park MD).

[68] This is but the briefest of a thumbnail sketch of the economic history described fully in Miller, *Way of Death*, chps. 2 and 3. For its expression in Gold Coast History, see Kea, "Trade, State Formation, and Warfare," 53, 57; and George Metcalf, "Microcosm of Why Africans Sold Slaves: Akan Consumption Patterns in the 1770s," *Journal of African History* 28, 3 (1987), 377–94.

[69] Van Dantzig and Jones, trans. and eds., *Pieter de Marees*, 58.

labors ashore. They directed the labor they appropriated towards clearing new fields on which their subjects sowed, tended, and harvested the commercially valuable grain.[70] Nobles extracted this unpaid labor from their subjects by claiming that all land belonged to them alone—an idea heard in the maxim: *ɔfum yɛ me deɛ, asase yɛ ɔhene deɛ* (Akan, 'the garden belongs to me but the land belongs to the chief')[71]—and they thus required that their subjects make and tend royal gardens before they could begin to develop and maintain their own plots.[72]

From the seventeenth century, noblemen diversified their trading affairs to include establishing plantations of food crops—especially maize—tended by slaves whom they accumulated as wealth.[73] They established these slave plantations in order to exploit the exploding demand for foodstuffs presented by growing coastal towns populated by non-farming traders, soldiers, and craftspeople to whom they could trade foodstuffs against imported commercial goods that they could then convert in the domestic economy for the slaves and other human dependants they primarily sought.

---

[70] Van Dantzig and Jones, trans. and eds., *Pieter de Marees*, 97; Hair, Jones, and Law, trans. and eds., *Barbot on Guinea*, vol. 2, 534; Kea, "Trade, State Formation, and Warfare,' 63.

[71] Wilks, "State of the Akan and the Akan States," 242. I have slightly altered Wilks's translation.

[72] Van Dantzig and Jones, trans. and eds., *Pieter de Marees*, 110–11; Kea, "Trade, State Formation, and Warfare', 53–54. Extended discussion of the legal (in the formal sense) underpinnings of these prerogatives are in K. A. Busia, *The Position of the Chief in the Modern Political System of Ashanti* (London: Oxford University Press, 1951), 18–50; and in formal-structural anthropological terms (again, as 'laws,' but in the cultural sense) in R. S. Rattray, *Ashanti Law and Constitution* (Oxford: Clarendon Press, 1929), 224–27, and M. Fortes, *Kinship and the Social Order* (Chicago: Aldine, 1969), 148–50. I take notice of the argument found in colonial-era literature that the idea promulgated by chiefs that their ownership of land was inalienable may have been a historically specific response to colonial encroachment on lands officials deemed 'unused' lands but were indeed left fallow to regenerate; see Sara S. Berry, *Chiefs Know Their Boundaries: Essays on Property, Power, and the Past in Asante, 1896–1996* (Portsmouth NH: Heinemann, 2001), 28n16.

[73] For secondary literature on the slave ownership by coastal élites, see, for the central coast, Larry Yarak, "West African Coastal Slavery in the Nineteenth Century: The Case of the Afro-European Slaveowners of Elmina," *Ethnohistory* 36 (1989), 44–60; and Adam Jones, "Female Slave-Owners of the Gold Coast: Just a Matter of Money?," in Stephan Palmié, ed., *Slave Cultures and the Cultures of Slavery* (Knoxville: University of Tennessee, 1995), 100–11. For the coastal towns of the south-east, see Ray Kea, "Plantations and Labor in the South-East Gold Coast from the Eighteenth to Mid-Nineteenth Century," in Robin Law, ed., *From Slave Trade to Legitimate Commerce* (Cambridge: Cambridge University Press, 1995), 119–43; and John Parker, "The Cultural Politics of Death and Burial in Early Colonial Accra," in David M. Anderson and Richard Rathbone, eds., *Africa's Urban Past* (Portsmouth NH: Heinemann / Oxford: James Currey, 2000), 205–21.

Maize planters recognized the vulnerability of their European customers, who were almost completely dependant on a reliable intake of local foodstuffs in order to sustain themselves both physically and commercially. Coastal grain suppliers exploited increasing competition in the seventeenth century among European factors to supply their forts with maize by soliciting factors to make material contributions at every major stage in the agricultural calendar. They invited the Europeans to send alcohol, livestock, and trade goods before their dependents would clear any land or sow any seed and, months later, harvest the fields.[74] Nobles could use the proceeds of such transactions to enhance ceremonies performed at the beginning of the agricultural season to assure good returns, thus strengthening their magico-cultural role in bringing forth rains to nourish the sown crops and assure successful harvests.

Planters strengthened their hand in negotiating final prices for maize delivery by acquiring the harvests of farmers working small-scale, privately-held plots they worked with their family dependents. The merchants' consolidation of maize supply constricted competition among grain-sellers and enabled them to get whatever price they demanded.[75] They also deployed their intimate knowledge of local market conditions and quick awareness of developments in fast-changing regional trade against their foreign customers' often obtuse notions about local affairs (and these slowed by bureaucratic management of intra-coastal messaging) in order to extract the highest possible prices for their sustaining commodity. With the increase in slave exports from the 1750s, Fante-speaking middlemen intensified their demands for tribute from the European factors they provisioned. They demanded imported luxuries with which they could distinguish themselves as 'big men' and could redistribute to stitch to together networks of belonging, including dependency. English merchants perhaps borrowed from their bilingual interlocutors when they recorded that the African merchants were 'eating' on the foreigners' perceived largesse.[76]

---

74 Multiple such transactions from the seventeenth and eighteenth century by English factors to notables in Fetu, Asebu, Fante, and Komenda are recorded in BNA T/70–accounts cited with full detail in Kea, "Trade, State Formation, and Warfare," 53, 54 with notes 2 and 3; and in Law, *The English in West Africa*, vol. 3, docs. 729, 790, 813. Similar Dutch gifts to the king of Fetu in 1645–46 are seen in Ratelband, *Vijf Dagregisters*, 31, 236.

75 Daaku, *Trade and Politics on the Gold Coast*, 121.

76 Ty M. Reese, "Eating Luxury: Fante Middlemen, British Goods, and Changing Dependencies on the Gold Coast," *The William and Mary Quarterly* 66, 4 (2009), 851–72.

In addition to owning slaves, rich and influential persons distinguished themselves from less fortunate people through their conspicuous consumption of expensive and esteemed edibles; People captured this perception in a proverb, *Osikani na odi ade a ɛyɛ dɛ.* (Akan: 'A rich man eats delicious foods.').[77] The wealthiest indigenous élites appropriated the culinary habits of high-ranking foreigners present on the coast and used them for their own purposes. They ate foods that were exotic, rare, and expensive.[78] This local phenomenon paralleled that of their similarly privileged contemporaries on the European mainland who too were embracing new, exotic foods which marked their integration into the commercialized global economy (principally spices and fruits, but also popular were those foreign edibles endowed with purported aphrodisiac powers).[79] Elite merchants of the Gold Coast consumed imported alcoholic drinks, preserved fruits and meats that only they could afford. They also spent resources on the conspicuous consumption of the most expensive of domestic foods—notably beef.[80] Cattle were rare, especially on the central Gold Coast, where endemic fly-borne disease made keeping them year-round untenable. In this era, cattle-herders of the dry grasslands of the south-east coast expanded their stocks, a portion of which they drove overland—and, apparently, sometimes using more extravagant and expensive (in terms of manpower and loss-risk in transit) canoe-transport—annually westward where they could resell them consecutively through local merchants until they could be presented, on the hoof, to rich merchants with the wherewithal to purchase a cow for public slaughter and consumption.[81]

---

[77] Christaller (compiler), *Three Thousand Six Hundred Proverbs*, trans. Kofi Ron Lange, 87.

[78] See above, chp. 3.

[79] E.g., bananas, which are frequently seen in contemporary European paintings where they were used as (phallic) symbols of lascivious sexuality; European visitors to West Africa drew a causal connection between banana consumption and sexual aggressiveness in women they encountered; see Van Dantzig and Jones, trans. and eds., *Pieter de Marees*, 162 with note 5. Also considered an aphrodisiac was the African gray parrot, which was not an uncommon part of European visitors' personal purchases while in Atlantic Africa—sometimes for resale to well-heeled and amorous gentlemen upon arrival at their home ports; see La Fleur, trans. and ed., *Van den Broecke's Journal*, 102.

[80] Van Dantzig and Jones, trans. and eds., *Pieter de Marees*, 42, 94–95, 168–69; Müller in Jones, trans. and ed., *German Sources*, 209, 239.

[81] Cattle traders reared calves on the plains on either side of the Volta mouth for sale to customers to the west; see Van Dantzig and Jones, trans. and eds., *Pieter de Marees*, 84; Müller in Jones, trans. and ed., *German Sources*, 140, 239; Winsnes, trans. and ed., *Erick Tilleman*, 29–30; Hair, Law, and Jones, trans. and eds., *Barbot on Guinea*, 434, 438, 484.

Wealthy men of the central coast organized themselves into a 'fraternity,' the signature public spectacle of which demonstrated their wealth and power in the cultural vocabulary of the new, Atlantic-era foodways. They paraded a cow festooned with garlands of maize ears (which were their source of their success), and the beast was covered in the paint powder (Akan, *kyire*) whose white color (Akan, *fufuu*) symbolized spiritual purity, victory, and joy.[82]

In the mid-seventeenth century the farmers, fishers, and traders of the Gold Coast littoral considered such cattle-slaughter and consumption to be a public demonstration of privilege.[83] Most of them could only occasionally afford to buy and consume poultry (which largely scratched out their own meager subsistence without tending, fattening themselves on crawling creatures and grains dropped during winnowing and threshing), and they considered the flesh of shorthaired sheep and goats tended by herders to be a rare delicacy.[84]

In contrast, poorer people of all nations subsisted on monotonous diets dominated by maize-based starchy foods—primarily *kenkey*[85]—which they ate with sauces (or 'soup' in modern, local parlance) flavored by vegetables and some flesh from less-esteemed creatures. Their sauces were probably most often shellfish based,[86] because these marine catches were extremely perishable and therefore inexpensive compared to fishes that could be preserved through smoking and drying in the sun. Coastal traders could reserve these value-added processed catches to trade inland against gold, slaves, and forest foods.

Prosperous persons could also distinguish themselves in private, by eating in distinctively foreign ways. When George Howland, an American captain, visited with Elmina merchants in the 1820s, he noted: 'They live in good style, having rich chased silver plates and dishes and silver and gold handled knives and forks and spoons. Their dinner consists of several courses, soup, meats, chickens, and fish, all well cooked, and fruits, nuts,

---

[82] Van Dantzig and Jones, trans. and eds., *Pieter de Marees*, 167–70. On the symbolism of color, see G. P. Hagan, "A Note on Akan Colour Symbolism," *Research Review* 7, 1 (1970), 8–14; for the description of the symbolism of white, see McCaskie, *State and Society*, 282.

[83] Van Dantzig and Jones, trans. and eds., *Pieter de Marees*, 42, 94–95; Müller in Jones, trans. and ed., *German Sources*, 209, 239.

[84] Müller in Jones, trans. and ed., *German Sources*, 209, 239.

[85] Cf. Kea, *Settlements, Trade, and Politics*, 302, which argued that yams were the food of the urban poor of the Gold Coast.

[86] This observation comes from the archaeological record, as presented in DeCorse, "Culture Contact, Continuity, and Change", 187–88.

and sweet-meats for desserts, and choice wines, licquors, and porter for drink.'[87]

### *Insecurity and Impoverishment Amid Scarcity and Violence*

When faced with dire shortages of food, family heads and merchants of the Gold Coast released some of their dependents for sale. Doing so was certainly not exceptional. Indeed, anecdotal evidence of selling relatives, especially children, into slavery amid conditions of drought and famine abounds in the early documentation of all of Atlantic Africa.[88] Though contemporary records of the Gold Coast indicate recurring episodes of scarcity following droughts, and situations in which scarcity contributed to slavery and other forms of dependency, it remains difficult to establish a direct causal relationship between adverse climactic events or trends and these human responses. Certainly it is difficult to see a correlation between records of drought and scarcity [see Illustration 4] and the volume of slaves exported from the Gold Coast [see Illustration 5].

First, there is a paucity of scientific data on climate variability and its human repercussions. This lacuna is gravest for the intra-annual phenomena of late or failed rains, which were the critical determinant of farming success in Africa. Furthermore, most scientific reconstruction of historic African climatology concerns the broader regional (or even continental) and not the extremely local context in which Gold Coast people experienced erratic conditions of food production and supply.[89]

Secondly, those who lived on the Gold Coast responded to scarcity in multiple ways, including seeking to mitigate hunger through social mechanisms of sharing or developing specialization.[90] Alternate strategies used scarcity to consolidate power through allocating scarce food resources to those desperate enough to accept social dependency as the price of asylum from hunger.[91]

---

[87] Norman R. Bennet and George. E. Brooks, eds., *New England Merchants in Africa: A History Through Documents, 1802 to 1865* (Boston: African Studies Program, Boston University, 1965), 118, cited in DeCorse, "Culture Contact, Continuity, and Change," 189.

[88] For West Africa, John D. Fage, "Slaves and Society in Western Africa, c. 1445–1700," *Journal of African History* 21, 3 (1980), 289–310, and La Fleur, trans. and ed., *Van den Broecke's Journal*, 28. For Central, Southern, and East Africa, see Miller, "The Significance of Drought, Disease, and Famine," 28 with note 56.

[89] James C. McCann, "Climate and Causation in African History," *International Journal of African Historical Studies* 32, 2–3 (1999), 261–79.

[90] McIntosh, "The Pulse Model."

[91] Miller, "The Significance of Drought, Disease, and Famine"; and Eldredge, "Sources of Conflict in Southern Africa." The theoretical orientation is from Sen, *Poverty and Famines*.

Illustration 4. Hunger and Disease on the Gold Coast, 1639–1919

| Date | Place | Event | Source |
|---|---|---|---|
| 1639–45 | Axim, Elmina | famine; 'At the same time there were many locusts, which ate their fruits in the fields, so that in many countries several hundred Moors actually died. They gave their fetissos much gold to drive them away, but it was in vain.' | Hemmersam in Jones, *German Sources*, 108 |
| 1640 | Accra | famine: 'bad harvest again at Accra from which it is feared much will perish . . . It appears that the traders are impoverished through famine . . .' | Arent Jacobsen van Amersfort (cited in Quaye, "Ga and Their Neighbors," 88) |
| 1645 | Fetu, other areas of Central Gold Coast | outbreak of disease (possibly diphtheria) claims the lives of trade slaves and many coastal residents (both local and European) | Jacob Ruychaver in Ratelband, *Vijf Dagregisters*, 93 |
| 1661, June | Central Gold Coast | famine: 'At present theare is noe Trade all over the Coast by reason of an extraordinary famine, people as they travelled dyed in the high way by theire goods, provisions not procureable for moneys. . . .' | Makepeace, *Correspondence of the English East India Company*, 96–7 |
| 1661 | Gold Coast; coastal areas east of the Volta | severe but short drought | Miller, "Unwanted Cargoes," 79; Greene, *History of the Anlo-Ewe*, 54. |
| 1679 | Accra | famine resulting from wars which keep farmers away from inland gardens | Barbot, 446n15 |
| 1681 | Adangme | smallpox outbreak among slaves | Law, *English in West Africa*, vol. 1, doc. 575 |

Illustration 4 (*cont.*)

| Date | Place | Event | Source |
|---|---|---|---|
| 1681, late March | Cape Coast | smallpox outbreak (at the castle only?) | Law, *English in West Africa*, vol. 1, doc. 220 |
| 1681, early May | Winneba | smallpox outbreak among slaves | Law, *English in West Africa*, vol. 1, doc. 587 |
| 1682, mid-to-late | coast east of the Volta | drought | Greene, *History of the Anlo-Ewe*, 54. |
| 1682–83 | Accra | famine | Miller, "Unwanted Cargoes," 79 |
| 1683, mid | coast east of the Volta | drought | Greene, *History of the Anlo-Ewe*, 54. |
| 1686 | Elmina | town decimated by smallpox | Claridge, vol. 1, 155, 165 (note: source is presumably Dapper, so the editor may have inferred the date from publication date of the translation/edition he consulted) |
| 1687 | Allada (Slave Coast) | famine, due to failed rains; many slaves sold | Van Dantzig, trans. and ed., *Dutch and the Guinea Coast*, 30–31. |
| 1692, April | Anomabu | no maize as a result of crop failure (following too much rain) | Law, *English in West Africa*, vol. 3, doc. 660 |
| 1693, February | Dixcove | food shortages: 'since I hear they [=the Africans] are soe poor that they would dayly pawne their children for money to buy themselves corne.' | Law, *English in West Africa*, vol. 3, doc. 23 |
| 1695, April | Komenda | 'no graine or corne and any fisch to be bought' so slaves forage for food | Law, *English in West Africa*, vol. 3, doc. 327 |

Illustration 4 (*cont.*)

| Date | Place | Event | Source |
|---|---|---|---|
| 1696 | Gold Coast | possible famine | Miller, "Unwanted Cargoes," 80 |
| 1700 | Accra | severe provisioning problems; trade slaves at Christianborg perish for want of food | Nørregård, 64 |
| 1703 | Accra | severe provisioning problems; trade slaves 'were dying by the day' | Justesen, *Danish Sources*, 124. |
| 1704 | SE Gold Coast | smallpox outbreak among trade slaves | Nørregård, 66 |
| 1719, March to November | Gold Coast | severe outbreak of unknown debilitating (fever, headache, weakness) and often deadly disease at Dutch forts and on ships; attributed to severe drought | OWIC 104 folio 231, Butler to Heren X, 27 March 1719 |
| 1721, January–June | Western Gold Coast | severe shortage of food due to war | NBKG 88, Willem Butler to Jacob Muller (Axim), 11 January 1721; Jacob Muller (Axim) to Elmina, 13 January 1721; Butler to Jacobus van Goch (Shama), 22 January 1721; Jacobus van Gogh (Shama) to Elmina, 25 May 1721 |
| 1721, January–October | Central Gold Coast | difficulty provisioning following Fante invasion of Fetu; compounded by drought; only provisions available are imported | NBKG 88, Willem Butler, 17 May 1721; Willem Butler to Louis de la Forterie and Jan Gottfired (Cape Coast) 23 June 1721; Kea, "Trade, State Formation," 193 |

Illustration 4 (*cont.*)

| Date | Place | Event | Source |
|---|---|---|---|
| 1722, February | Accra | observation from Danish governor at Chistianborg that 'many slaves grow ill at the fort because of the bad food, which consists of two *millie* loaves and some salt water per day' | Justesen, *Danish Sources*, 287 |
| 1723 | Accra | impossible to buy provisions due to war with Akwamu | Nørregård, 74 |
| 1724, March | Accra | scarcity of food because Akwamu sacked (in 1723) Aprag and Soko, which were Accra's granary | European dispatch dated 21 March 1724 (cited in Quaye, 133) |
| 1724 | Axim | extreme scarcity of food | TWIC 106 folio 35 Pieter Valckenier to Heren X, 20 July 1724 |
| 1739 | coast east of the Volta | drought | Greene, *History of the Anlo-Ewe*, 54 |
| 1740 | Accra | famine | Reindorf, 277 |
| 1743 | Gold Coast | food shortages | Miller, "Unwanted Cargoes," 82 |
| 1747, May | Accra | violence destroying inland areas where maize is grown, so some of Danish company's enslaved people 'have starved to death.' Also, conspiracy to revolt on outward-bound ship due to poor provisioning. | Justesen, *Danish Sources*, 698, 711 |
| 1747, August | Accra | At Christianborg, enslaved people die due to underfeeding and by taking their own lives | Justesen, *Danish Sources*, 700 |

Illustration 4 (*cont.*)

| Date | Place | Event | Source |
|---|---|---|---|
| 1747 | Accra | for two months, slaves at Christianborg forage for food in order to sustain themselves | Justesen, *Danish Sources*, 711 |
| 1748 | coast east of the Volta | drought | Greene, *History of the Anlo-Ewe*, 54 |
| 1754 | Gold Coast | Asante war with Denkyura, Twifo, Wassa and Akyem threatens the region with famine | Thornton, *Warfare*, 68 |
| 1758 | coastal areas east of the Volta | coastal flooding | Greene, *History of the Anlo-Ewe*, 54 |
| 1764 | Akyem | Asante war against Akyem threatens the area with famine | Thornton, *Warfare*, 72 |
| 1766 | central coast | famine | Reese, "Eating Luxury," 864 |
| 1774 | Gold Coast | period of starvation | Miller, "Unwanted Cargoes," 83 |
| 1774 | SE Gold Coast | high mortality on slaves ships due to food shortages at disembarkation points on the Gold Coast | Nørregård, 128 |
| 1774 | coast east of the Volta | drought | Greene, *History of the Anlo-Ewe*, 54 |
| 1780 | coast east of the Volta | drought | Greene, *History of the Anlo-Ewe*, 54 |
| 1780s | West Africa | regional drought; hunger in Akuapem | Miller, "Unwanted Cargoes," 83; Isert 161 on Akuapem |
| 1780, late–1781 mid | Accra | impossible to buy provisions | Nørregård, 80–81 |

Illustration 4 (*cont.*)

| Date | Place | Event | Source |
|---|---|---|---|
| 1792 | SE Gold Coast | Popo soldiers conquer Anlo, when the latter run out of food | Thornton, *Warfare*, 68 |
| 1795 | Gold Coast | measles and smallpox | Miller, "Unwanted Cargoes," 84 |
| 1807 | Accra | famine | Reindorf, 276 |
| 1809 | coast east of the Volta | drought | Greene, *History of the Anlo-Ewe*, 54 |
| 1800s (?) | Asante; Annomabo | smallpox outbreak | Meredith, 161 |
| 1809 | Accra | Asante invasion causes famine | Reindorf, 276 |
| 1811, October | Akyem | smallpox breaks out among Akyem soldiers | Cruickshank, 97 |
| 1816 | Accra | famine following Asante occupation of farms on the outskirts of Accra, and which prevented sowing of seeds to grow in the 'small' rainy season of 1815 | Reindorf, 276; Kea, "Some Danish Accounts," 6; Nørregård, 192 |
| 1816 | Central Coast | 'thousands die' during famine 'induced by the restrictions put on cultivation by the Asante invasion, as well as by their consumption of the fruits of the earth.' | Cruickshank, 128 |
| 1817 | Asante | skin disease in slaves | Bowdich, 375 |
| 1822 | Accra | famine | Reindorf, 276 |
| 1824 | Gold Coast, particularly Accra | failed rains cause crop failures | Miller, "Unwanted Cargoes," 85; Nørregård, 195 |

Illustration 4 (*cont.*)

| Date | Place | Event | Source |
|---|---|---|---|
| 1824 | coast east of the Volta | drought | Greene, *History of the Anlo-Ewe*, 54 |
| 1824, July | Asante | hunger caused by late onset of rains; and smallpox outbreak among Asante soldiers | Cruickshank, 160; Wilks, *Asante*, 180 |
| 1825 | Accra | famine | Reindorf, 276 |
| 1829 | Accra | famine | Reindorf, 276 |
| 1832 | Accra | famine | Reindorf, 276 |
| 1833 | Accra | locust infestation, but damage limited | Reindorf, 277 |
| 1834 | Accra | locust infestation | Nørregård, 211 with 256n36 |
| 1835 | Accra and Akuapem | drought brings crop failure | Nørregård, 211 with 256n36 |
| 1836 | Accra and Akuapem | drought brings crop failure | Nørregård, 211 with 256n36 |
| 1836–39 | Accra and Akuapem | famine; locust plague arrives in May 1838 and destroys all of the maize, yam, and cassava. Locusts followed by cancer-worm infestation. Pests finally leave by February 1839. | Reindorf, 277; Nørregård, 211 with 256n36. |
| 1842 | Accra and Akuapem | locusts return | Reindorf, 278–78 |
| 1845 | Accra and Akuapem | drought brings crop failure | Nørregård, 211 with 256n36 |
| 1846 | Accra and Akuapem | drought brings crop failure | Nørregård, 211 with 256n36 |
| 1852 | coast east of the Volta | drought | Greene, *History of the Anlo-Ewe*, 54 |

Illustration 4 (*cont.*)

| Date | Place | Event | Source |
|---|---|---|---|
| 1862, July 10 | Accra | earthquake destroys much of Accra | Van Dantzig, *Forts and Castles*, v; Claridge, vol. 1, 501 |
| 1871 | Elmina area | smallpox epidemic begins late in the year and lasts for several years | Baesjou, 42, 210, 212 |
| 1873, June | central coast | war refugees already emaciated and sickened crowd towns of the central coast, where there is little food; many die of starvation and smallpox outbreak rages | Claridge, vol. 2, 27, 33–35 |
| 1880s | Asante | five years of civil war brings collapse of food farming, and regional famine follows | McCaskie, *State and Society*, 29 |
| 1891–93 | Gold Coast | influenza pandemic | Patterson, 1995 |
| 1893 | coast east of the Volta | drought | Greene, *History of the Anlo-Ewe*, 54 |
| 1904 | Accra | several years of drought force residents to abandon gardens near the town; severe hardship for urban and rural dwellers | Parker, *Making the Town*, 170 |
| 1908, January–May | Accra | plague outbreak | Claridge, vol. 2, 575 |
| 1910 | Sekondi | yellow fever outbreak | Claridge, vol. 2, 575 |
| 1918–19 | Gold Coast | influenza pandemic | Patterson, 1995 |

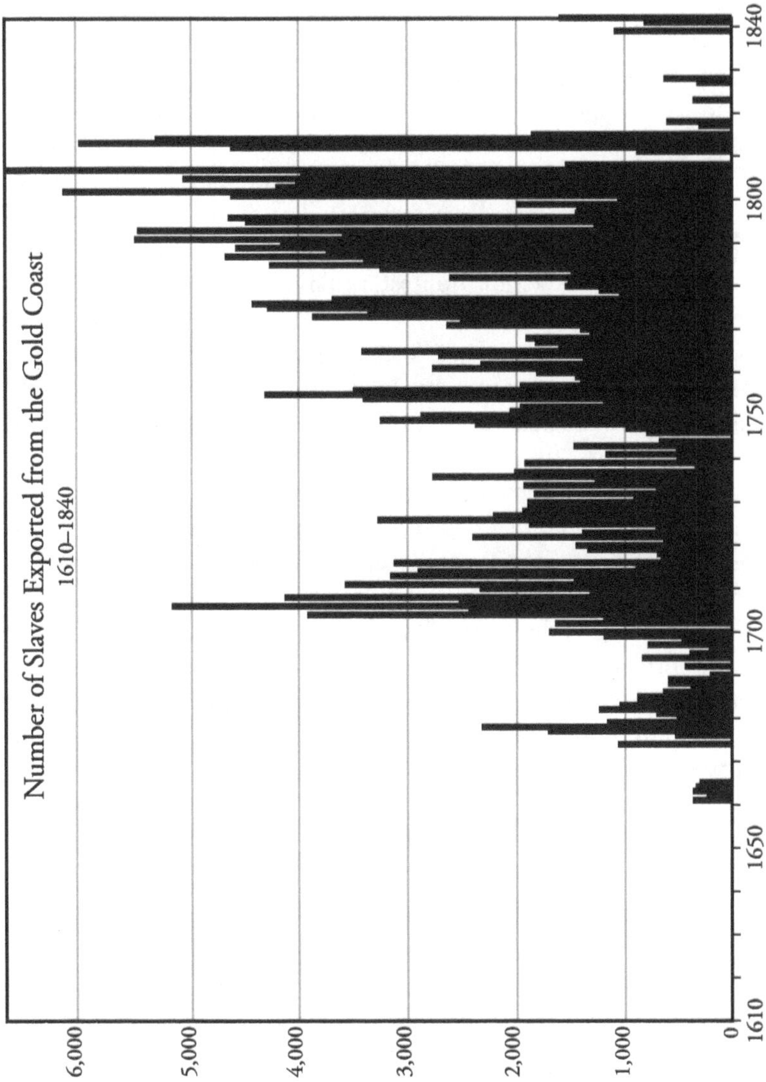


Illustration 5. Volume of Slaves Exported from the Gold Coast
(source: slavevoyages.org)

Nor was the sale of slaves, even by family members, always the regrettable response of extreme desperation. Sometimes it found place within the context of household politics and personal ambitions.[92]

Finally, the complexity of these human responses to scarcity means that significance of the climatic events and subsequent scarcity are not to be found solely in the annual numbers of slaves exported. The greatest significance of the hunger and hardships might rather be the expansion of domestic slavery and other forms of dependency within the Gold Coast itself, where indeed most of those removed from their natal communities in the Atlantic era remained in bondage.[93] And many, perhaps most, of those who became enslaved were exported as a cumulative result of pawnship that could not or would not be redeemed by their families or masters, release for sale within domestic markets for slaves, and finally purchase by European merchants.

Slave-holding African merchants keen to the cool logic of preventing loss of perishable inventories of living beings released persons they could not sustain with food in exchange for durable trade goods that they could convert later back into human wealth when edibles were again abundant. European buyers on the coast entered the same concerns into the corollary calculus of acquiring the greatest number of slaves possible to be had at low prices in terms of food scarcity and its consequent abundance of slave supply while operating within the limits of their own ability to provision, if but minimally, these persons while warehoused in the cellars of their coastal forts and transported below-deck across the Atlantic. European factors tried to use their superior regional transportation and logistics (and, perhaps, purchasing power of imports) to acquire food beyond the local area(s) suffering food shortages at a given moment in order to acquire greater numbers of slaves for export. But, failing their ability to acquire food locally or regionally, they halted all purchases of slaves in order to mitigate financial ruin.

---

92 Charles Piot, "Of Slaves and the Gift: Kabre Sale of Kin During the Era of the Slave Trade," *Journal of African History* 37, 1 (1996), 31–49. For a narrative example, see Miller, *Way of Death*, 3.

93 This hidden sector of the Atlantic diaspora of African slaves is highlighted by Pier M. Larson, *History and Memory in the Age of Enslavement: Becoming Merina in Highland Madagascar, 1770–1822* (Portsmouth NH: Heinemann, 2000), ii, 268–82; and Robin Law and Paul Lovejoy, "The Changing Dimension of African History: Reappropriating the Diaspora," in Simon McGrath, *et al.*, *Rethinking the African Diaspora* (Edinburgh: University of Edinburgh CAS, 1997), 181–200.

Not all famines were induced solely by unavoidable pestilence and failed rains. Armies of Gold Coast polities created famines as a matter of military tactics. Mobilized troops lived 'off the land,' feeding themselves by plundering the fields of their enemies. Such plundering was certainly the most economic manner in which to wage war, not only because commandeered foodstuffs came without payment but also because maintaining logistical systems to provision armies on the move would have necessarily relied on head-portage, which was labor-intensive and therefore expensive. Military commanders therefore timed their maneuvers just after the final rains of the season, when they could exploit gardens and fields filled with mature produce ready to be harvested.[94]

Mobilized troops plundered while on the move, but they could also set up more permanent camps on fertile fields in order to further weaken their enemies. Foreign military occupation of ripening fields also threatened farmers at their most vulnerable moment, in the 'hunger season,' when their stores of the previous season's crops were depleted but the earliest of the current crops were just maturing. Occupation through the harvesting season compounded farmers' problems by preventing them from reaping the fruits of their previous agricultural labors. These challenges were intensified if enemy troop encampments remained in their fields, as this deprived farmers from preparing gardens for the planting of the next season's crop. For example, when Asante armies invaded the south-east Gold Coast in 1815–16:

> [the Asante soldiers] pitched their camp right on the Accra Negroes' farming placed called Onianse, about two miles from Accra and this town [=Christianborg, a.k.a. Osu]. They are still there and eat up all of the coastal Negroes' corn, yams, cassava, *gobbegobbess* (a kind of bean),[95] *pessiange* [=*pisang*, i.e.,

[94] As Tilleman noted: '[war] occurs most after the *travat* times [=rainy season; Portuguese, *trovoada*, 'a hard, sudden storm'], in the months of June and July here, when the fruits begin to ripen, so that they have food everywhere'; Winsnes, trans. and ed., *Erick Tilleman*, 30. Such timing also afforded easier troop movement across rivers retreating from the engorged state characteristic of waterways in the rainy seasons.

[95] These beans were probably the Bambara groundnut (*Vigna subterranea* [formerly classified *Voandzeia subterranea*]); see Hepper, *The West African Herbaria of Isert and Thonning*, 101, 168. This local vernacular, *gobbegobbe*, derived from one of the Bantu languages spoken in the area of the Congo River, in which such groundnuts were known by reflexes of *nguba*. In the Atlantic-era speakers of many of the Bantu languages saw such similarity between the Bambara Groundnut and the new, American groundnut, or peanut (*Arachis hypogaea*) that they applied the name of the familiar crop to the foreign one. Their recycled term for peanuts finds clear echoes in the speech of the American South, where peanuts are known as 'goobers.' The Anglophone term 'peanut' similarly derives from Atlantic-era African speech. Speakers of Bantu languages from north of the Congo

> bananas] and *bacotiti* [=*bachoven*, i.e., plantains] which they had sowed and planted for this year's provisions [...] In about four months from now we will all come to a time of hunger for bread [and] *titi* [=porridge],[96] because the planting time for... the whites' New Year harvest [i.e., in January, at the end of the 'small' rainy season] is nearly over and no coastal negro has dared to go to plant and sow as long as the Assiantees lay encamped in their corn fields.[97]

Farmers and those who depended on their produce were equally reluctant to leave the relative safety of their villages to do the extensive canvassing necessary to find wild, famine foods they might forage, or to hunt for game, as they rightly felt unprotected there from violence, capture, and enslavement. In such desperate circumstances, people turned to eating anything available, such as weedy grasses and the starchy underground parts of wild plants that they normally considered unfit for human consumption and that they could render edible only with much extra work.[98]

### *Impoverishment in Times of Plenty*

Eras of bountiful maize harvests held equally serious, but hidden, dangers for those dependent on this foreign grain. They suffered from severe malnutrition even though their bellies were full. Such 'hidden malnutrition' was the inevitable outcome of a maize-dominant diet, because this grain presented a particularly poor nutritional profile to human consumers.

---

referred to the old, and then new, sorts of groundnuts as reflexes of *pinda*. In the American South, English-speakers who overheard their slaves use this term to describe what was at the time only an important subsistence for slaves coined the new term 'peanut' by conflating the African sounds they heard with their own folk terms for similar foods (pea + nut) they knew. On the Gold Coast, farmers everywhere cultivated peanuts, but only as a minor crop, as was the case with the Bambara groundnut; Hepper, *The West African Herbaria of Isert and Thonning*, 84, 101.

[96] Gã, *jíjí*. Although now cooks primarily make their porridge with cassava flour, formerly they used grain or specific plantain cultivars that were particularly suitable (like the one apparently identified in the passage above; *bacotiti* < *bachoven* + *jíjí*). Captives from this region brought this style of porridge and its name (as '*gigi*') to the plantations of Suriname, and then into the bush when they ran away, where they used roasted and ground maize mixed with plantains. See Price, "Subsistence on the Plantation Periphery: Crops, Cooking, and Labour among Eighteenth-Century Suriname Maroons".

[97] R. A. Kea, "Some Danish Accounts of the 1815–16 Asante Military Expedition on the Gold Coast," *Ghana Studies Council Newsletter* 10 (Spring/Summer 1997), 6.

[98] Abbiw, *Useful Plants of Ghana*, 30–31, 41–43, 210, 216–17. For a contemporary account, see Rømer's account of Accra residents 'eating grass' in Winsnes, trans. and ed., *A Reliable Account of the Coast of Guinea*, 224 with note 116.

While no grain offers an abundant amount of protein, the deficiency is the most pronounced in maize,[99] and half of its protein is in a chemical form that is not easily digested. Furthermore, maize lacks sufficient amounts of the important amino acid known as lysine, which is needed to metabolize protein. Maize grain is also terribly deficient in niacin (or nicotinic acid, commonly known as vitamin $B_3$), and most of this is chemically bound to another molecule that makes it unavailable to its (single-stomached) human consumers. Moreover, maize is lacking in the amino acid known as tryptophan, which is needed to metabolize niacin. Thus those who depend on maize for their primary food risk acquiring serious vitamin and protein deficiency diseases.

In the wake of the Columbian exchange and the introduction of maize to the Old World, nutritional deficiency disease stalked those who made the new grain their primary starch. In Mesoamerica, whence maize was introduced to the Old World, cooks had developed sophisticated processing techniques that allowed them to place maize foods at the center of the foodways without exposing maize eaters to malnutrition. They combined dried maize grains and water, to which they added slacked lime ($CaOH_2$; available in deposits of natural sodas or rendered from burnt seashells to which water was added), and brought this mixture to a boil. After an hour or two of simmering, they removed the mixture from the heat and let it stand for another half- or full-day. This process loosened maize grains' pericarp, which cooks then washed away along with any excess lime. Cooks then ground the mash into the fresh dough known to the Spaniards as *masa* that was then turned into starchy staple foods such as *tortillas* and *tamales*.[100] This dough boasted increased amounts of nutritionally available niacin and a much better balance of protein to essential amino acids. But maize seeds arrived in Europe and Africa unaccompanied by this critical culinary technology.

The sharecroppers of southern Europe, like the farmers of the Gold Coast, sowed maize because they appreciated its great productivity. The Spanish poor were accordingly the first community in Europe to suffer from maize-related nutritional deficiency diseases, in the first decades of

[99] This lack of protein was remarked upon earlier, in chapter 3, in reference to its culinary significance: that maize's lack of gluten—strings of protein, which trap gasses inside loaves—made it unsuitable for making risen breads.

[100] In the mid-twentieth century, innovations in industrial food processing enabled factories to dry and pulverize *masa* to make the dry, stable, and storable maize 'flour' known as *masa harina*, which is now a familiar product on the shelves of American supermarkets and, in Europe, those of some ethnic groceries.

the eighteenth century, and the new illness was named *mal de la Rosa* (lit., 'pink sickness') for the discoloration of the irritated skin of the afflicted. The modern name for the disease was coined by an Italian scientist in 1771, who noted that *pellagra* (lit., 'rough skin') was this disease's most immediately conspicuous symptom.[101] Yet the identification of pellagra's specific aetiology in niacin deficiency went unidentified until the 1930s; until then, speculations ranged widely—but tended to blame the victims for their poor habits, 'breeding' (i.e., genetics), and so forth.

As early as 1600 pellagra ravaged those who made the new grain their primary starchy food on the Gold Coast. At that time, the Dutch visitor De Marees observed that: '[t]hose who eat much of it [=maize], because they find it delicious, but without being used to it, tend to be tortured with scabs and itch, or at any rate are plagued with great boils, because it makes the blood hot.'[102] This disease was certainly pellagra, and the serious skin eruptions he noticed indicate an advanced stage of the disease.[103] De Marees's observations of these symptoms are certainly acute and are, apparently, the heretofore unrecognized first European record of the disease anywhere. His remarks are all the more extraordinary for his correct identification of a maize-dominant diet of as the culprit.

Pellagrins suffered other symptoms too, of which De Marees was apparently only partially aware. As he noted, the common diet of the Gold Coast in the early Atlantic era resulted in 'indigestion' among both Europeans and indigenous people;[104] gastrointestinal disturbances such as nausea, vomiting, and diarrhea are common symptoms of the early stages of pellagra sickness. And perhaps the psychological symptoms of the disease—including insomnia, depression, rapid mood swings, dementia—were

---

101 McGee, *On Food and Cooking*, 242–44; Daphne A. Roe, *A Plague of Corn: A Social History of Pellagra* (Ithaca: Cornell University Press, 1973).

102 Van Dantzig and Jones, trans. and eds., *Pieter de Marees*, 113 (=Naber, ed., *Goud Koninkrijck*, 119 with note 1). For contemporary European notions of 'heat' in diets, see above, chapter 3, note 11.

103 *Contra* the editorial comments of Van Dantzig and Jones, trans. and eds., *Pieter de Marees*, 113n2: 'It is difficult to see how eating maize could cause itches, scabies, or boils.'

104 Van Dantzig and Jones, trans. and eds., *Pieter de Marees*, 43. This modern edition leaves De Marees's use of the now obsolete Dutch word *ternisse* untranslated, with an explanatory note suggesting weakness, irritability, and a hungry feeling. I translate here with thanks to René Baesjou (in conversation at my house in Leiden in 1999) who was confident that the De Marees's meaning was 'indigestion' when I instead initially offered that it might denote 'waste away' or 'look pale' from the French verb *ternir* (perhaps due to the author's familiarity with the language). The Dutch and French words do share a history. The *Woordenboek der Nederlandsche Taal* gives 'indigestion' as an accepted but atypical definition, and cites De Marees as the first reference.

beyond his purview or were otherwise subsumed under the rubric of his description of decidedly non-European customs and behaviors of those with whom he kept company on the coast. Pellagrins unable to resolve their malnutrition eventually die.

The incidence of pellagra on the Atlantic-era Gold Coast must have shadowed the great success of maize cultivation on the coastal grasslands. The favorable environmental conditions, which accelerated successful experimentation with the new grain in the early sixteenth century, continued for at least a century to bless maize farmers with bountiful harvests. At the time of De Marees's visit around 1600, maize was abundant and failure of this very drought-sensitive crop was not recounted.[105]

The protein deficiency in maize-dominant diets preyed upon the weakest members of Gold Coast communities: children, whose growth was stunted by the protein deficiency disease known as kwashiorkor.[106] Mothers who removed their nursing children from the breast and put them onto the soft adult food—notably, maize porridge—they could eat frequently saw these weanlings fail to thrive. Instead, the weanlings became extremely inactive, apathetic, and irritable and showed pronounced swellings and reddish tint to their skin and hair. Some went into shock and coma before perishing. Others who recovered never recouped normal physical nor mental growth and agility. Moreover, malnutrition increased the vulnerability of growing children to infections—both endemic (like malaria) and opportunistic (which lodged in respiratory, intestinal, and skin systems) resulting in disease, added energy loss, and shared misery and impoverishment.[107]

Gã- and Dangme-speaking mothers named this condition *kwashiɔkɔ*,[108] which signaled their identification of the precise social conditions of the disease's onset[109]—when an elder child became sick (Gã, *kwá*, 'to fade

[105] Van Dantzig and Jones, trans. and eds., *Pieter de Marees*, 44, 110.

[106] Recently, world health campaigners have included this disease under the umbrella of 'Protein-Energy Malnutrition.' But the thesis that protein deficiency is the underlying cause of the disease does have its critics; see Michael H. N. Golden, "Oedematous Malnutrition," *British Medical Bulletin* 54, 2 (1998), 434–44.

[107] Ulrich E. Schaible, and Stefan H. E. Kaufmann, "Malnutrition and Infection: Complex Mechanisms and Global Impacts," *Public Library of Science* 4, 5 (May 2007), 806–12.

[108] Or, kwashiorkor, as it is written in the orthography of the European languages of medicine and world health programs.

[109] Their understanding of the disease as arising within a social context is consistent with other Africans' approach to health and illness; see John Janzen and Steven Feierman, eds., *The Social Basis of Health and Healing in Africa* (Berkeley: University of California Press, 1992).

away') once s/he was removed (Gã, *shí*, 'to leave aside') from the mother's breast in order to accommodate a newborn sibling, or if the child was orphaned.[110] These mothers, like mothers everywhere, then put their weanlings onto the soft adult food they could eat. In the south-east of the Gold Coast, these were porridges of boiled and mashed grain—especially maize. Children reared in the moister climes of the Gold Coast mostly escaped this disease, because the dominant starchy staples there—plantains, rice, or yams—with which parents could make weaning foods lacked the hidden protein deficiency of a maize diet.[111]

Nutritional deficiency diseases that developed amid eras of plenty afflicted individuals and burdened their social networks in ways akin to those wrought by hunger and malnourishment in times of war and bad harvests. For individual victims of deficiency diseases, falling chronically ill created great personal physical suffering and psychological anguish in the short term. Permanently diminished physical strength and agility and compromised mental acuity and stability impoverished the afflicted in the long term, as these disabilities left them ill-equipped to reap the benefits of all subsequent good agricultural years and unable to calculate and execute complicated agricultural strategies needed to meet the challenge of making a living amid less favorable environmental conditions.

Victims therefore burdened their kin and larger communities of belonging with recurring needs for assistance with food, shelter, and even companionship—as those victims who escaped death but suffered lasting, significant psychological troubles made for particularly poor marriage candidates. Such psychological problems were understood in local terms as arising from witchcraft and malevolent spirits.[112] Survivors of nutritional

[110] An alternative etymology would draw on the Gã day-name *Kwashi*, for a male child born on a Sunday (Cf. Akan, *Kwesi*).

[111] C. D. Williams, "A Nutritional Disease of Childhood Associated with a Maize Diet," *Archives of Disease in Childhood* 8, 3 (1933), 423–33; C. D. Williams, "Kwashiorkor, a nutritional disease in children associated with a maize diet," *The Lancet* 266 (1935), 1151–52; F. I. D. Konotey-Ahulu, "Kwashiorkor," *British Medical Journal* 302 (1991), 180–81; F. I. D. Konotey-Ahulu, "Issues in Kwashiorkor," *The Lancet* 343 (1994), 548; and personal communication with Dr. Konotey-Ahulu.

[112] During Cicely Williams's pioneering research (see note above), she observed that mothers and families were reluctant to share their knowledge of the disease, sought 'traditional' healers, and sometimes abandoned sick children for the sake of the family. For an account of Williams's research in the place of gender and the history of medicine, see Jennifer Stanton, "Listening to the Ga: Cicely Williams' Discovery of Kwashiorkor on the Gold Coast," in Lawrence Conrad, and Anne Hardy, eds., *Women and Modern Medicine* (Amsterdam: Rodopi, 2001), 149–171 [special issue of *Clio Medica: Perspectives in Medical Humanities* 6 (2001)].

deficiency who remained unattached deprived their families of the opportunity to make social alliances that might be used to buffer hardships. Malnutrition thus impoverished both individuals and their kin.

## *Conclusion*

Slaves and other free but junior dependants ultimately bore the consequences of increasingly risky, short-term commercial strategies of kings, 'big men,' merchants to pursue subsistence strategies based on new, starchy crops. In the forest, plantain farmers found themselves the subject of increasingly onerous forms of social dependency and economic predation, and victims of violent struggles to control the gold trade. On the coastal grassland, maize farmers suffered during both boom and bust, through compromised health, impoverishment, hunger, debt, and enslavement as rains needed to sustain this grain crop became increasingly variable in the eighteenth century.

CHAPTER FIVE

# THE PORCUPINE'S SHAME: BEARING THE BURDEN OF CASSAVA CULTURE

Gold Coast farmers did not begin their experiments with cassava cultivation and consumption until the middle of the Atlantic era, long after they had made the other, new foreign starchy staples—maize, plantains, and Asian rice—their own. Indeed, most Gold Coasters did not cultivate or consume much cassava until the very end of the Atlantic era, after the closing of the (legal) trans-Atlantic slave trade in the early nineteenth century. Gold Coast farmers' early experiments with cassava cultivation demonstrated to them that this tuber was a foul-tasting food with possible dangers, owing to the plant's indigenous potential to poison those who ate it.[1] They integrated cassava into their food strategies as a hunger-food only after a long period of botanical modification through which they altered the original, poisonous types of the introduced plant into an indigenous crop with diminished dangers. Even as late as the end of the eighteenth century, cassava cultivation and consumption were limited to coastal areas prone to droughts and crop failures, during which cassava was eaten as a famine food. The adoption of cassava as a staple crop throughout the region was finally sparked in the mid-nineteenth century by, first, the arrival of Africans from the New World diaspora who knew techniques to process the poisonous roots into a desirable food and who invested in cassava cultivation and processing on Gold Coast plantations to satisfy their own cravings for food from their displaced homes in the diaspora; and second, early nineteenth century drought and warfare that

[1] All varieties of cassava (*Manihot esculenta* Crantz) present powerful toxins related to cyanide which are, obviously, injurious to human health. Of greatest concern are the cyanogenic glucosides (which yield prussic acid, or HCN, by digestive decomposition) which are present in all parts of this plant. Botanists (and Africanist historians who borrowed from them) formerly recognized two distinct species of cassava which were classified according to their toxicity, with 'bitter cassava' (*M. utilissima* Pohl.) and 'sweet cassava' (*M. palmata* Muell.). However, because levels of toxicity are a relative measure and because certain ecological conditions (including soil nutrient deficiency and drought) are known to have a strong influence towards increasing toxicity regardless of a individual plant's genetic breeding, the bifurcated classification scheme is unsatisfactory.

created famine conditions in which cassava was seen as a reliable producer of bulk food to fill empty bellies.

### *Problems in the Earliest Records of Introduction*

The earliest three references—one Dutch, one English, and one Danish—cited in modern historical literature[2] as describing the earliest cultivation of cassava on the Gold Coast do not adequately indicate local adoption into farming regimes or culinary repertoires.

The earliest possible Dutch-language report dates from 1683, when the directors of the Brandenburg African Company[3] established rules that would govern the operation of the fort, Gross-Friedrichsburg, they were then establishing just west of Cape Three Points. These directions drew distinctions between esteemed, imported European foods (wheat bread and bacon were specifically named) that were to be reserved for officers' fare, and locally available foods to be used for the garrison and, presumably, for the slaves they hoped to acquire. These latter groups, the directive reads, were, 'to be given fish and *carsauen* or local bread and rice.'[4] As Adam Jones, the editor of the modern, critical edition of Brandeburg correspondence noted, '[t]he use of the term *carsauen* is interesting, although it does not in itself prove that cassava was being cultivated on the Gold Coast by this time.'[5] Not only is the identity of '*carsauen*' ambiguous, but, further, even if it were not and the document were to read "cassava" (or another, recognized, contemporary name in Dutch for cassava) the historical weight given to such company correspondence from European headquarters to African posts on local African matters would be slight. In such missives, the authors of the information had perhaps no primary experience in the region themselves and perhaps were conflating

[2] Alpern, "The European Introduction of Crops to West Africa," 164n169.

[3] The Brandenburg African Company (1682–1720), although established in Germany, employed many Dutchmen, hence the use of Dutch language in much of the company's correspondence.

[4] Adam Jones, trans. and ed., *Brandenburg Sources for West African History, 1680–1700* (Stuttgart: Steiner, 1985), 76, 252. The phrasing in the Dutch original ('*carsauen off broodt van het landt*...') seems to indicate an appositive construction, which would supply two descriptions of the same item. Thus the use of '*carsauen*' would indicate that the author was making direct reference to the realities of provisioning in the West Indies, where '*carsauen*' [=cassava] was the local name for bread made from processed cassava root; see below.

[5] Jones, trans. and ed., *Brandenburg Sources*, 76n6.

knowledge about other tropical locales, including the West Indies or West-Central Africa, where cassava was indeed common provision for the poor and enslaved.

The British source for cassava cultivation regards a fort garden experiment that seemingly was not sustained. In a third-person précis of a letter Sir Dalby Thomas wrote in 1706 from Cape Coast castle to the Royal African Company, it was reported that, 'Has Cassado continuing to thrive.' Dalby was known to be interested in raising crops successful in the West Indies in hopes of improving the diets of Englishmen and exploring the possibilities of commercial, non-food crops in West Africa. He had two acres planted with indigo, which he reported was faring well and so hoped to expand to twenty acres in the following year. Among food crops, there were peas, 'french Beans,' and this 'Cassado.'[6] However, there is no contemporary evidence of indigenous Africans in the area integrating any of these crops into their gardens.

The earliest Danish source for cassava on the Gold Coast is found in the travelogue authored by Rev. Johannes Rask, who visited the area in 1708–9. After Rask's death in 1744, his son prepared his father's manuscript into a posthumous publication in 1754.[7] Stanley Alpern, the historical cataloger of foreign foods in West Africa, initially cited this account as describing that cassava was 'commonly grown in the Accra area' and with 'its root made into flour for bread' there, but corrected this in a revision of his catalog of exotic plants.[8] The particular portion of Rask's travelogue which concerns cassava at Accra[9] finds its place not in the long, descriptive prose about food crops grown on Gold Coast, but rather in logbook-like daily entries which recount Rask's experiences after his Gold Coast visit, on his homeward bound voyage via Annobón. On this island, as Rask's account attests in passages of text both previous and subsequent to that about Accra, contemporary locals indeed grew cassava and processed its starchy roots into 'bread.' Rask's paragraph about Accra, however, makes it very clear that

[6] Initially cited by Jones, trans. and ed., *Brandenburg Sources*, 76n6. The citation is BNA T70/5 Thomas to Royal African Company, 16 June 1706. I thank Stan Alpern for sharing a copy and his historical insights with me.

[7] Rev. Johannes Rask, *En kort og sandferdig Rejse-beskrivelse til of fra Guinea* (Trondheim, 1754). Because the original publication seems to be rather rare, historians have frequently used the more common modern edition, which is a translation of the original Danish into New Norwegian: Jotein Øvrelid, ed. and trans., *Ferd til of fra Guinea* (Oslo: Fonna Forlag, 1969).

[8] Alpern, "The European Introduction of Crops to West Africa"; Alpern, "Exotic Plants of Western Africa".

[9] In the Øvrelid edition, this is the last paragraph of p. 159.

*no* cassava was eaten there, beyond a perhaps exaggerated account of ethnopharmacological use of cassava root:

> This plant, cassava, is known everywhere around Christiansborg Castle, but one never sees any animal eating it, not of any part, either leaves or stem. This makes me believe even more strongly in the things the Negroes say about cassava, which has leaves like hops, but not as sharp as hops leaves. The root stock, or tuber, is reddish and thick, like a large cabbage stalk. The Negroes say that if you have a toothache, you can just cut a small piece of the root, as long and wide as the area of the painful teeth, and let it lie along the gum for half an hour, then the teeth will loosen so you can pluck them out with your fingers.[10] On these cassava bushes grow beautiful black, clear [?translucent] berries, and I offered some of these to a monkey, which is the surest way to find out if something is edible. I would advise everyone to stay away from unknown fruits that the monkeys avoid, but this monkey shoved the cassava berries away and shrieked. I conclude from this that there must be a ruinous and poisonous juice in them.[11]

Rask's book then proceeds to describe the 'cassava bread' he ate while at Annobón.

The text's confusing insertion of information about Christianborg cassava into a section of about Annobón foodways and the modern text's difficult language[12] leaves room for understandable confusion. There remains the possibility that the organizational misplacement was created by Rask's son-*cum*-editor as he rearranged parts of the original manuscript for publication; such movement of manuscript material between long descriptive passages and daily journal entries while editing an early African travelogue for publication appears to have been a not uncommon practice in the Rask's era.[13] Examining Rask's account for evidence of cassava cultivation and consumption underscores the difficulty in reading early texts for West African history, and the care scholars must bring to the task.

Sources contemporary to Rask's visit to Accra (c. 1708) appear to support—albeit through negative evidence—the interpretation put forward

[10] Unlikely; even the tubers of the most bitter types of cassava do not contain enough poison to dislodge a living tooth. Such use of cassava root is certainly not part of modern Ghanaian folk medicine, which does have a broad collection of natural remedies for toothache; see Abbiw, *Useful Plants of Ghana*, 192–94.

[11] Øvrelid, ed. and trans., *Ferd til of fra Guinea*, 159. I thank Selena Winsnes for discussing this source with me (personal communications, 25–26 April 2001).

[12] New Norwegian is very unusual research language for Africanists and, indeed, this recently 'invented' language is challenging to many modern Norwegians.

[13] For discussion of this phenomenon with regards to the sole surviving Dutch prepublication manuscript, see La Fleur, trans. and ed., *Pieter van den Broecke's Journal*, 13–17.

here: that cassava was not cultivated as a food crop at that time. Erick Tilleman, a Dane who authored the most descriptive account of agricultural practices in the area around the fort at Christiansborg, recorded no cassava cultivation or consumption there circa 1697.[14] Likewise, Willem Bosman's comprehensive account of Gold Coast, based on his unusually long experience working there, circa 1688–1702, lacks a single mention of cassava cultivation or consumption anywhere on the Gold Coast, including at Accra.[15]

## *Introducing Cassava*

What Rask's account makes clear, however, is that although Danes and Africans near the Christiansborg fort were not cultivating cassava for food circa 1708, the plant was indeed growing there at the time. Identification of Christiansborg castle as the (so far as yet known) initial point of origin for Gold Coast cassava points to two possible European agents of introduction. Probably it was either the Danish or Portuguese agents who alternated control of this fort[16] and were thus the first to plant cassava near Christainborg. In 1661, the Danish West India Company had made payments to the local ruler, one 'Kanckoy, King of Great Accra' in order to secure permission to build the fort from which they soon thereafter commenced trade, initially as a satellite point of the main Danish position at Cape Coast but later, in 1685, as the headquarters of the Danish Company in Guinea.[17] Perhaps it was one of those experimentally minded Danish commanders of the early eighteenth century[18] who initiated the first attempts to grow cassava on the Gold Coast in the same manner as

[14] Winsnes, trans. and ed., *A Short and Simple Account of the Country of Guinea and Its Nature* (Madison: African Studies Program, University of Wisconsin Press, 1994).

[15] Bosman, *Nauwkeurige beschryving van de Guinese Goud- Tand- en Slave-Kust.*

[16] In addition to the Europeans who alternated in control of the fort, an Akwamu merchant-chief captured the fort through trickery in 1693 (he had formerly been a cook at the castle), and then held it temporarily before the Danes repurchased control of it the following year; Van Dantzig, *Forts and Castles*, 31–32; Daaku, *Trade and Politics on the Gold Coast*, 111–12.

[17] The document is included (via a German-language copy in the Danish archives) in Ole Justesen, ed., *Danish Sources for the History of Ghana 1657–1754*, 2 vols. (Copenhagen: The Royal Danish Academy of Sciences and Letters, 2005), vol. 1, doc. I.5.

[18] Several are known to have imported botanical resources with which they experimented locally; Georg Nørregård, *Danish Settlements in West Africa, 1658–1850* (Boston: Boston University Press, 1966), 72, 249n4.

it was produced to feed slaves in similarly pelagic Danish dominions in the West Indies.[19]

Or perhaps this stand of cassava was the product of Portuguese experimentation at Christiansborg during a brief period of Portuguese possession of the fort. In 1679 a former governor of São Tomé bought the fort from a group of Danish Company mutineers who had successfully seized control of the property. Although these Portuguese kept the fort for less than five years before commercial losses[20] forced them to accede to a Danish bid to re-inhabit the fort in 1683,[21] the São Tomé Portuguese may have, in the interim, introduced cassava from their island home where this crop had been a cultivated starchy crop at least since mid-century.[22]

All slave traders based at Accra lodges faced daunting challenges in finding food in sufficient quantities to maintain their perishable inventories of people, not to mention themselves. European traders on the Gold Coast had always therefore regarded a posting to Accra as something of a hardship. Although this area was not infrequently the source of good trade in gold and slaves, the area's marginal agricultural production persistently threatened personnel with physical and commercial ruin. Poor rains or the violent disruption of starch sales from inland areas (principally nearby Akuapem) could bring Europeans there to the brink of starvation and block the departure of ships. As early as the mid-seventeenth century, Dutch merchants described their operations at Accra as wholly dependent on outside food.[23]

---

[19] Although European planters (primarily Dutchmen expelled from Barbados) succeeded in developing sugar plantations on the Danish island of St. Croix, similar efforts to develop commercial agriculture on St. Jan and (particularly) St. Thomas failed because of the arid conditions, poor soils, and difficult topography. On the importance of cassava in these islands' slaves' diets, see Winsnes, trans. and ed., *Paul Erdmann Isert's Letters*, 185.

[20] The Portugese enterprise at Osu was hamstrung by their inability to compete with English and Dutch traders who offered higher quality goods just a short distance up the coast, at James Town (known locally as Nleshi) and Ussher Town (or Kinka), respectively.

[21] Van Dantzig, *Forts and Castles*, 31–32.

[22] São Tomé colonists appear to have pursued cassava cultivation from the mid-seventeenth century for domestic consumption as well as for export as a processed good for provisioning, in the form of flour (Port., *farinha*), to be traded against mainland slaves and European merchandise; see Adam Jones, trans. and ed., *West Africa in the Mid-Seventeenth Century: An Anonymous Dutch Manuscript* (Atlanta: ASA Press, 1995), doc. 6. A contemporary Dutch source, however, remarked that while some cassava was grown on the island, yams were instead the main starchy staple and little cassava was actually eaten there; Arnout Leers, *Pertinente beschryvinghe van Africa* ... (Rotterdam, 1665), 317–18.

[23] Dapper, *Naukeurige beschrijvinge der Afrikaensche gewesten*, vol. 2, 82. Dapper's description of the Gold Coast draws on information from the 1640s, despite its later publication

The most successful European companies invested significant resources in creating and maintaining logistical systems to provision Accra forts with food purchased by factors at windward ports from African entrepreneurs.[24] These organizational efforts included dispatching small European craft loaded with food, but more often they employed those African middlemen known on the Gold Coast as *rimadoors* (from the Portuguese *remeiros*, 'rowers'; alt. Port. *remador*) who shuttled edibles on large canoes better suited than were sailed craft operated by Europeans to travel short-distances between Gold Coast ports, close to shore in shallow waters, and against prevailing winds.[25] However, not even the best-run provisioning systems could alleviate hunger in Accra if drought scorched the entire region or if marginal harvests produced only slim surpluses too expensive for European factors to buy in sufficient quantities and ship to Accra.

But, in this era, Christiansborg-based traders lacked logistical systems of similar sophistication and reliability to counter localized food shortages. The São Tomé Portuguese traders were but a low-priority interest operating within a larger imperial scheme that was chronically deficient in shipping capacity and organizational integration. The Danish traders found themselves similarly neglected by their metropolitan backers while the home country was consumed with wars against Sweden (1700–1720).[26] Christiansborg slave traders who were unable to secure adequate provisions for their slave inventories saw potential profits wane into debt balances when hungry slaves succumbed to starvation or died from diseases that preyed upon the malnourished. In the face of such increasingly grave provisioning problems, Christiansborg-based traders probably introduced cassava cultivation to the Gold Coast in hopes of stimulating local food production which could alleviate their exposure to financial ruin like those created when slaves held at the fort died for want of food.[27] This

---

date; see Adam Jones, "Decompiling Dapper: a Preliminary Search for Evidence," *History in Africa* 17 (1990), 171–209.

24 See chapter 4.

25 Josée Nan, "Rimadoors: een Afrikaans aandeel in de handel tussen Nederland en de Goudkust, 1602–1872" (unpublished thesis, Universiteit Leiden, 1981). Sailed ships were at a particular disadvantage in attempting to make a westward return voyage from Accra. A Dutch source put the time required for a European ship to sail from Accra to Elmina against the prevailing eastward winds and current at some three weeks; see Jan Simon Gerhardus Gramberg, *Schetsen van Afrikas Westkust* (Amsterdam, 1861), 21–22. Rimadoors could cover the same distance in a third of the time.

26 Nørregård, *Danish Settlements in West Africa*, 72.

27 For specific examples, Illustration 4 in the previous chapter; and further narrative description in Nørregård, *Danish Settlements in West Africa*, 64, 66, 128.

crop-introduction strategy was identical to that pursued by the earliest Portuguese traders on the Gold Coast, who introduced the first batch of foreign starches—plantains, maize, and Asian rice—two hundred years before. European slavers disembarking from ports of the southeastern Gold Coast, however, do not appear to have ever in this era taken cassava tubers onboard for provision during transport, given the extreme perishability of the raw tuber once unearthed.[28] Once out of the ground, cassava tubers will spoil within a week.

After the early European experiments with cassava cultivation failed to attract local African interest, Osu residents and fort personnel presumably left this stand of cassava abandoned, to fend for itself against infertile soils, arid conditions, and seasonally-brisk sea winds.[29] Local observers must have noticed that this foreign plant thrived nonetheless, and plants that achieved maturity produced seeds encapsulated in fruits—like the ones Rask reported, and which never are seen on cassava plants used in an intensive cultivation regime[30]—and that these seeds produced new plants.

## *Africanizing Cassava Culture*

Although indigenous farmers thus must have found that cassava plants flourished in the arid climate and infertile soils of the Osu environs of the Danish fort Christiansborg, African consumers considered this plant's parts to be largely unpalatable. Tasters found that its leaves made for poor, bitter-tasting spinach and its starchy tubers were foul-tasting even after they had been processed according to familiar techniques used on yams (e.g., peeling and boiling for *ampesi*; further pounding for *fufu*). Furthermore, they found that the tubers had an unwelcome tendency to become fibrous, or 'stringy,' particularly when the tubers were grown to achieve maximum size, which made *ampesi* unpleasantly chewy rather than

[28] For a representative provision list, see Per Hernæs, *Slaves, Danes, and African Coast Society: The Danish Slave Trade from West Africa and Afro-Danish Relations on the Eighteenth Century Gold Coast* (Trondheim: Department of History, University of Trondheim, 1995), 340–41.

[29] Although cassava is native to sheltered rainforest environments, it can prosper in vastly different conditions; see Irvine, *West African Crops*, 154–55.

[30] Farmers pursuing intensive strategies harvest young plants at the point when they have tubers of acceptable size but before they attain sexual maturity, which induces changes in the plant's tubers, including a tendency towards fibrous growth and bitterness; E. V. Doku, *Cassava in Ghana* (Accra: Ghana Universities Press, 1969), 6–7.

tender and also made the work of women to pound the cooked starch into *fufu* all the more arduous.[31] The Gã-speaking experimenters of Osu named the trial crop *duade*, which perhaps captured their perception of the plant as a cultigen (*dũ̀*, 'to plant') but one that produced a marginally edible and certainly not ideal raw food (*adɛɛ*, 'small bad things').[32]

Experimenters were not blind to the potential of cassava however. They saw that it produced bulk starch when their other locally grown starchy crops (principally maize and millet, and some yams) failed due to poor rains, plant disease, or pests. Furthermore they discovered they could harvest cassava tubers at any time of the year, including in the hungry season before the principal starch crops were ready for harvest. Experimenters therefore began a long process of trying to adapt the introduced varieties of cassava they found near Christiansborg towards new types that exhibited characteristics they desired. They sought to ameliorate the qualities of the crop they saw as negative, principally the high poison content as recognized by taste, while maintaining attractive features, such as hardiness.[33]

Nineteenth-century oral sources remembered that these experimenters found the process extremely difficult and that they almost abandoned the project for lack of progress.[34] Experimenters faced a steep learning curve in understanding how to develop the crop towards their preferences through selective reproduction. That mature cassava plants produced seeds was obvious, but because these plants were *highly* sexual once they reached maturity, experimenters were unable to isolate specific plants targeted for their favorable characteristics from cross-pollination by plants with undesirable attributes. And although cassava roots resembled yam tubers, experimenters who replanted the 'head' of the cassava tubers in the same manner they propagated favored yam cultivars proved only that these cassava tuber cuttings rotted rather than sprouted.[35]

---

31 Doku, *Cassava in Ghana*, 6, 16.

32 If the Akan-speakers had named the plant during the brief Akwamu interregnum at Christiansborg, then a rough, literal translation of *duade* might therefore be, 'tree yam' (*dua* + *ode*) which is remarkably similar to the way contemporary Dutch and German sources identified cassava, as 'stick-yam' (*stokjam*). Note also that the Gã word *duade* (also *odɛɛ*) identifies the sea perch (*Petrometopon nigri*).

33 Doku, *Cassava in Ghana*, 16.

34 Reindorf, *History of the Gold Coast and Asante*, 267.

35 Doku, *Cassava in Ghana*, 4–17. Modern crop science has since largely eliminated the ability of commercial cassava cultivars to reproduce sexually. Unlike other edible tubers, cassava does not produce a wound response when sliced during its dormant stage.

The critical discovery for experimenters was that the woody stalks of cassava they cut down and discarded on the garden floor to serve as mulch would begin to root if there were sufficient moisture; moreover, they observed that these new plants exhibited all of the characteristics of the parent plant, because they were in fact genetic clones. With this discovery cassava developers could select specific plants with desirable characteristics and from them take multiple stem cuttings to nurture. And they could supplement their asexual development program by allowing some favored stem-grown cultivars to reproduce sexually and produce seeds that might produce plants with novel characteristics of interest. Experimenters monitored fallow cassava stands as a laboratory of feral genetic innovation.[36]

Cassava gardeners were also pleased to find that plants grown from cuttings produced tubers of harvestable size sooner—in most cases in half the time—than plants grown from seed. Cooks also learned that clones had an additional advantage over their seed-grown siblings, since tubers grown from a stem cutting produced tubers less fibrous and thus more suitable for making *ampesi* and *fufu*.[37]

Intermittent food shortages in the southeast throughout the mid-eighteenth century certainly gave experimenters the imperative to continue their development of cassava as a hunger food [see Illustration 4]. Although the women of Accra were probably critical to the local adaptation of this crop, as they had been with other, initially-foreign starches disparaged by men, whatever critical role played had been forgotten by the later nineteenth century. Then, oral traditions credited development of the new types to the intercession of a male spirit who washed away forever the toxins with his urine.[38]

By the 1780s farmers in the area of Accra had produced cassava cultivars that were distinct from original, introduced stock from the New World. Their cassava was less poisonous, and these highly developed Gold Coast

[36] Doku, *Cassava in Ghana*, 8–9; Dan Kojo Affran, "Cassava and Its Economic Importance," *Ghana Farmer* 12, 4 (1968), 172–73.

[37] Doku, *Cassava in Ghana*, 8–9; Affran, "Cassava and Its Economic Importance," 173–74; C. Annor-Frempong, "A Survey of Cassava Cultivation Practices in Ghana," *Acta horticulturæ* 380 (1994), 216–18.

[38] Reindorf recorded the tradition: 'As traditionally reported, their fetish Sakumo promised to pass its urine on the root so as to remove that power from it, which he did, and so it became good'; Reindorf, *History of the Gold Coast and Asante*, 267.

cultivars could be identified by their unique leaf color and shape.[39] Danish botanists working in the area recorded that, '[t]he Guinea plant differs from the [West] Indian in its smaller leaves, which above are darker, below nearly light blue; and in that the root is not poisonous.'[40] These types were however not completely free of cyanogenic glucosides, but cooks who peeled the tubers as a routine part of their basic culinary preparation made the remaining starch safe (perhaps unknowingly), because the latent toxicity of the developed, 'sweet' types resided in that outermost layer of starchy pulp attached to the fibrous husk they removed.

Accra consumers considered these new types to be edible even without any significant processing to remove remaining root poison. Thus when hunger gripped Accra residents in 1783 following poor harvests in Akuapem, they could fill their bellies with cassava tubers cooked in the least labor-demanding manner: they simply uprooted and placed unpeeled tubers on a fire to roast, and once the tubers were soft they skinned and ate them.[41] Accra residents did not prefer to eat cassava compared to other starches. Rather, as one long-serving British factor noted at the beginning of the nineteenth century, they ate cassava because all other starchy crops were prone to frequent failure there.[42]

The development by Accra plant breeders of less-bitter cassava cultivars prompted farmers in neighboring areas to borrow the crop and integrate cassava into their food-strategies as a fail-safe starch. Gã- and Dangme-speakers farming the lands between Accra and the village known as Pokoasi, located where the plains give way to forests and steep elevations, integrated cassava into their gardens by planting the new crop amongst their beans and other vegetables.[43] There, they found, cassava provided light shade to the tender leaves of seedlings and prevented the

---

[39] Winsnes, trans. and ed., *Paul Erdmann Isert's Letters on West Africa and the Slave Trade*, 185; Hepper, *The West African Herbaria of Isert and Thonning*, 57–58.

[40] Hepper, *The West African Herbaria of Isert and Thonning*, 57–58.

[41] Isert reported that at Accra, '[t]he Blacks prefer to eat these cassavas after they have been roasted in a fire, like chestnuts'; Winsnes, trans. and ed., *Paul Erdmann Isert's Letters on West Africa and the Slave Trade*, 125. Isert, in his travelogue, understated the problem of persistent hunger in the area. Perhaps he lent a blind eye to it, however, because describing the area accurately as agriculturally marginal would have generated doubts about the viability of Danish-backed plantations he hoped to establish there.

[42] Meredith, *An Account of the Gold Coast of Africa*, 190.

[43] Reindorf, *History of the Gold Coast and Asante*, 279. Several small villages inland of Accra whose gardens included such mixed cropping are named in De Marrée, *Reizen op en beschrijving van de Goudkust van Guinea*, vol. 1, 217.

sun from robbing light soils by evaporation of scarce moisture needed by maturing crops. These farmers might unearth cassava tubers as needed once the other crops had been harvested, but if they were not wanting for extra starch, they could leave cassava in the garden as it went to fallow. Here, mature cassava plants nurtured the new weeds that grew spontaneously once a garden was abandoned, thus accelerating the rejuvenation of the environment.[44]

These farmers' Ewe-speaking neighbors across the Volta River began to plant cassava on the light soils of the Volta Plains in the same manner, but they placed cassava at the center of their food strategies more rapidly to meet the demands of an extended period of extreme demographic growth and simultaneous environmental stress. From 1679 the farmers received the first wave of what would become several subsequent migrations of refugees fleeing military conflicts unfolding on the other side of the river.[45] Most refugees clustered for protection in coastal towns that were physically separated from the mainland violence by lagoons and mangrove swamps. The streams of refugees must have strained the capacity of the indigenous Anlo-Ewe-speaking farmers and lagoon fishers[46] to produce adequate supplies of edibles for the burgeoning population of the coastal towns. These farmers further recognized that only limited areas of the coast had the soils to sustain millet crops, and the limited soil nutrients on these plots could not sustain intensification by skipping the traditional fallow period and continually sowing the same plot with millet seed.

Additionally, farmers of the lower Volta Plains and the coastal strip who considered expansion of existing crops faced great risks due to the area's unreliable rainfall and the unsuitability of much of the area's grassland for millet farms. Here to the east, they suffered crop-ruining droughts more often than farmers on the other side of the great river. The area received less annual precipitation, and had no mountain ranges in the hinterland that might provoke passing moisture-laden clouds to give up

---

[44] H. Bramer, "Soils," in J. B. Wills, ed., *Agriculture and Land Use in Ghana* (London: Oxford University Press, 1962), 17.

[45] Sandra E. Greene, *Gender, Ethnicity, and Social Change on the Upper Slave Coast* (Portsmouth NH: Heinemann, 1996), 20, 26, 68–71.

[46] Although the Anlo-Ewe are now regarded to be master fishers of Ghanaian saltwater, their activities were until recent times confined to inland waters and labors ashore. See Emmanuel Akyeampong, *Between the Sea and the Lagoon: An Eco-Social History of the Anlo of Southeastern Ghana c. 1850 to Recent Times* (Athens OH: Ohio University Press / London: James Currey, 2001).

their rain. And when rains did fall, both of the predominant soil types farmers encountered (black clays inland and sandy soils near the coast) evacuated rainfall more quickly than the slightly heavier soils of Gold Coast savannas and forests.[47]

Through planting cassava, Anlo-Ewe farmers were able to intensify agricultural production of bulk starch within the strict environmental limits of their lands. The essential ingredient of this intensification was additional labor inputs, which the indigenous farmers recruited from amongst the war refugees. With the addition of cassava to their food strategies, these farmers were able not only to support a larger population but also to save some of their fish and cattle that they had previously sold inland for additional bulk carbohydrates and instead divert these protein-rich foods into making profits which they could invest into accumulating more dependents.

If cassava was initially the food of the poor, the displaced, and the indebted, it quickly became a marker of general ethnic identity. So important was cassava to the Ewe-speakers, that they named it, succinctly, 'there is life' (*agbeli*). Their great reliance on this crop stimulated cooks to use their creativity to experiment with new processing techniques and culinary styles in order to introduce variety into their diets. In addition to recipes such as *fufu* and *ampesi*, which were popular starchy food forms throughout the Gold Coast, Ewe cooks developed new styles based on cassava dough or *agbelima* (*agbeli* + *ma*, 'dough') which they fermented to improve the taste and nutrition of the starch. This food product also resisted spoilage for a few weeks, so cooks could prepare dough in batches, then remove smaller portions to use a little at a time, as needed.[48] They also created other starchy dishes, including *agbelikaklo* (fried cakes of grated cassava), *yakayake* (steamed cassava bread), and *akple* (a stiff, slightly fermented cassava paste), which they used to accompany soups made with vegetables, fish, and meats.[49] Nowhere else in the region did cooks in early times develop such a broad assortment of cassava recipes.

On the coastal grasslands west of Accra, Awutu- and Fante-speaking farmers sensed none of the urgency that encouraged Ewe farmers to make cassava their staple and palatable starch so quickly. Awutu-speakers near

---

[47] Papadokis, *Crop Ecologic Survey in West Africa*, map 23; Ofori-Sarpong, "Rainfall Reliability over Ghana"; Boateng, *Geography of Ghana*; 154.

[48] S. Sefa-Dedeh, "Process and Product Characteristics of *agbelima*, a Fermented Cassava Product," *Tropical Science* 35 (1995), 359–64; Doku, *Cassava in Ghana*, 33.

[49] Doku, *Cassava in Ghana*, 32–34; Field, "Gold Coast Food".

the European forts at Winneba borrowed the crop developed by Accra farmers (and the Gã name for it)[50] in the early nineteenth century as an experimental one and found multiple ways in which cassava stands supplemented their existing reliance on grains. They planted cassava around the periphery of gardens to create a living hedge which defined boundaries and repelled free-roaming livestock from encroaching on the enclosed maize and vegetable patches. Farmers of the coastal grasslands found that while cattle browsed on tender sprouts of the cassava hedge, the plant prospered nonetheless, without causing any great injury to the roots, which remained as a hunger food for farmers.[51]

Fante farmers also used cassava to guard against the return of irregular but well-remembered famines by planting the crop on lands, like beach cliffs, that were too dry, nutrient-poor, or otherwise unsuitable for maize (which did not tolerate saline spray) or plantains (which required shelter from wind, and moisture). In Fante proverbs, people poked fun at cassava as a food even the lowly porcupine was ashamed to eat, (Akan, *Okuafo afɛre ɔbankye, na kɔtɔkɔnso afɛre*: 'The farmer is ashamed to eat cassava and the porcupine is also ashamed to eat it.').[52] But they also ridiculed farmers who didn't plant some of the crop as insurance against the persistent threat of famine.

Indeed farmers of the central coast expanded cassava cultivation only after the Asante invasion in 1806–7 and the implementation of brutal policies of military occupation that lasted until 1826. Many farmers sought

---

[50] Gã *duade* > Awutu *juade*. Perhaps the crop and the name was introduced by cattlemen who drove their herds seasonally westward towards (European) markets on the central coast, while taking advantage of the fresh shoots of spring grasses; for a contemporary description on this trade in cattle, see above, chp. 4.

[51] P. van IJzendoorn, *De Goudkust: beknopt overzigt van de aangelegenheden betrekkelijk de kust van Guinea, en aldaar van de Goudkust in het bijzonder* ('s-Gravenhage and Amsterdam: Gebroeders Van Cleef, 1817), 18; Gramberg, *Schetsen van Afrikas Westkust*, 40. By making cassava thickets, farmers might have also been appealing the Fante notion that planting—especially trees—signaled settlement, and delineated the boundary between order (cleared and cultivated space) and disorder (unworked lands); see T. C. McCaskie, "Nananom Mpow of Mankessim: An Essay in Fante History," in David Henige and T. C. McCaskie, eds., *West African Economic and Social History: Studies in Memory of Marion Johnson* (Madison: African Studies Program, University of Wisconsin, 1990), 135.

[52] Christaller (compiler), *A Collection of Three Thousand and Six Hundred Tshi Proverbs* [1879], trans. Kofi Ron Lange, 244. Porcupines, as well as other ground mammals, do indeed sometimes rob gardens of cassava tubers, but only when they do not find other, better foods to purloin. But because the porcupine is a symbol of the Asante people, the proverb may also have another meaning; this would point to Asante culinary prejudice against cassava, which they consider(ed) to be suitable only for the poor, the misfortunate, and foreigners.

refuge during the first invasion by hiding in the forests, caves, and other agriculturally marginal areas[53] where they might have used quick-growing cassava as a reliable hunger-breaker. During occupation of the coast, 'many thousands' were reported to have died during a famine 'induced by the restrictions put on cultivation by the Asante invasion, as well as by their consumption of the fruits of the earth.'[54] In the plundered and occupied area, subsistence farmers gleaned survival from cassava roots that, though nutritionally deficient, were hidden safely underground and resisted spoiling there for several years.

Once coastal subsistence farmers were relieved of onerous Asante occupation, they reconstructed former food strategies but retained cassava in a more prominent role than before. They began to include it in their crop rotations, planting it after other, older staple crops, such as plantains, maize, and yams, had tired garden soils. Thus, these farmers reduced their need to organize male labor—scarce in the wake of wars and slaving—to clear new fields required for the preferred foodstuffs. On older, cleared fields, wives and daughters carried the burden of cassava production. They planted, tended, and harvested these crops and processed the roots into a palatable meal if it were needed. Although they told proverbs that explained how they no longer regarded cassava as the reviled food it had once been,[55] consumers on the central coast continued to prefer and celebrate all other cultivated starches before they resorted to eating the mocked root.[56]

In the deciduous forest north of the central coast, Asante farmers working lands outside of the capital of Kumase began to add substantial amounts of cassava to their crop rotations only after 1850. Farmers in areas such as Kwabre (north of Kumase) and Atwoma (to the southwest) were accomplished food producers whose bounty had long supplied the markets of Kumase and the kitchens of its urban, non-farming population with produce. But by the later nineteenth century, these farmers were confronted with declining soil fertility wrought by at least two centuries of intensive cultivation.[57] Through shortening the fallow period they could

---

53 Rev. Hans Christian Monrad, *Bidrag til en skildring af Guinea-kysten og dens Indbyggere* (Copenhagen: A. Seidelin, 1822), chp. 4.

54 Cruickshank, *Eighteen Years on the Gold Coast of Africa*, vol. 1, 128.

55 J. Yedu Bannerman, *Mfantse-Akan mbɛbuɛsm nkyerɛkyɛrmu* (Tema: Hacquason Press, n.d.), 226.

56 Gramberg, *Schetsen van Afrikas Westkust*, 39–40.

57 Abayie Boaten, "Asante: the Perception and the Utilization of the Environment before the Twentieth Century," 25.

continue to prepare gardens of relatively rich (if not technically 'virgin') soil on which they could sow maize seed and make yam mounds to produce their preferred starches, albeit with ever-declining returns. In the later nineteenth century, these farmers then experimented with cassava cultivation to extend the period of a particular garden's utility.

Their innovation was to plant cassava on the plots they would otherwise return to fallow once the preferred starches had depleted the soils of nutrients. Their experiments with this sort of cassava cultivation showed them that cassava flourished even on such degraded soils, thus producing a second crop of starch without having to expend the tremendous labor efforts to clear heavy vegetative cover for a new garden. They further recognized that this strategy allowed them longer fallow times to previously used soils, during which forb regrowth provided more nutrients in the form of ash when they again cleared the plot for the primary crop of yams or maize.[58]

The principal challenge for these farmers was to find to find a local market of consumers who were willing to eat foods derived from cassava. The market they found was among the communities of foreigners in Asante, counted among them newcomers, often slaves or the descendants of slaves originally from the coastal area, who were on the one hand economically disadvantaged and thus unable to afford high-prestige starches such as yams, and on the other hand were familiar with cassava consumption in their natal communities, where this crop had been grown and eaten.[59] It is significant that the principal cassava-growing areas, Kwabre and Atwoma, were the principal settlement areas for slaves captured in battle since the early eighteenth century.[60]

To the west of the Central Coast, along the coastal stretch of rainforest west of the Pra River known as Ahanta, farmers' upland gardens of yams and plantains were better and more dependably watered. Moreover, the heavily forested country impeded most slaving violence. Ahanta farmers probably began experimenting with cassava planting soon after their Fante-speaking neighbors to the east did, but linguistic evidence points to significant independence in their early crop trials [see Word List 9, page 192]. Ahanta farmers did not borrow their neighbors' term for

---

[58] Bramer, "Soils".

[59] Abayie Boaten, "Perception and Utilization of the Environment," 25.

[60] Abayie Boaten, "Kumase: Early Settlement up to the End of the 18th Century," 11; D. J. E. Maier, "Military Acquisition of Slaves in Asante," in Henige and McCaskie, eds., *West African Economic and Social History*, 119–32.

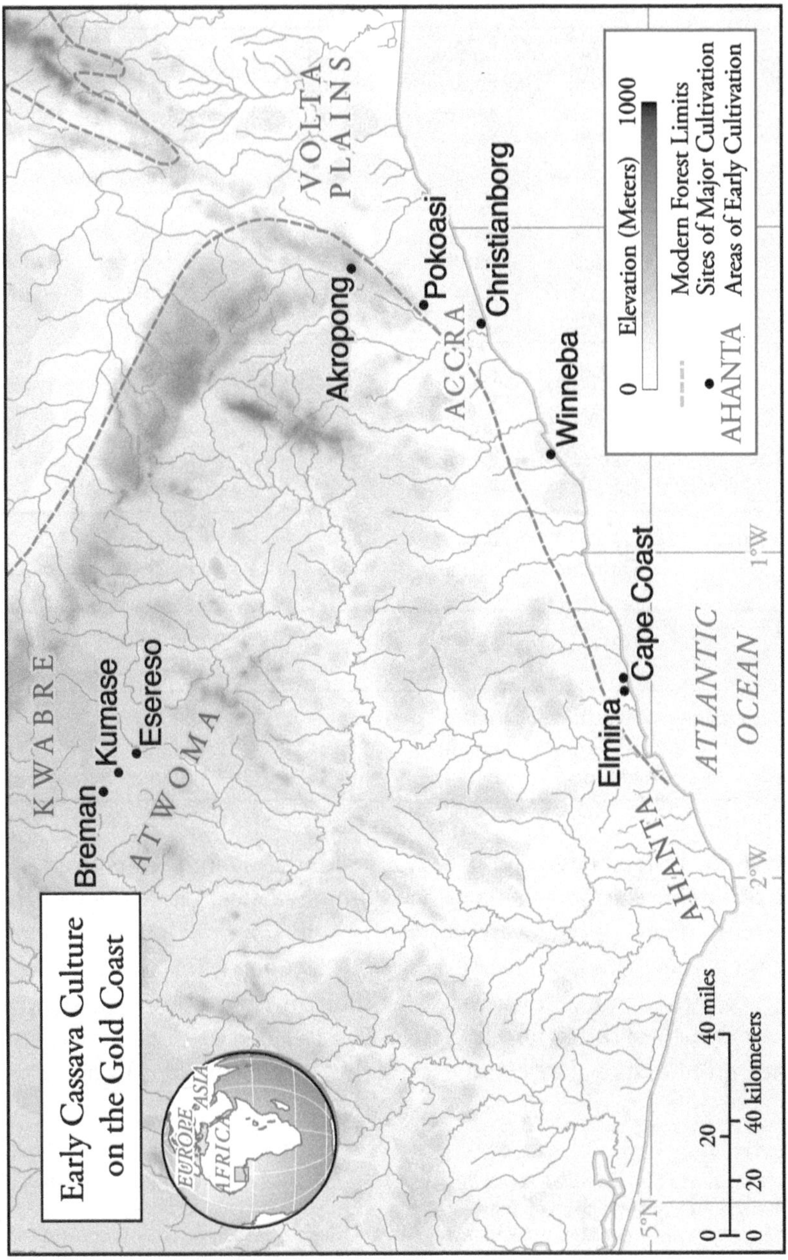


Map 9. Early Cassava Culture on the Gold Coast

Word List 9. Comparison of Vernacular Names for Cassava

| Language | Dialect | Cassava | Comment |
|---|---|---|---|
| Akan | Fante | bankye | |
| | Nzema | c. 1817 *abromenade* (De Marrée)<br>bɛdɛ | neologism: *boro*, 'overseas' + *awennade*, lit. 'to eat iron' (plant name) |
| | Akuapem | bankye | |
| | Asante | bankye | |
| | Brong | bankye<br>bɛdɛ (alt.) | <br>cf. Nzema *bɛdɛ* |
| | Chakosi | | |
| Guang | Awutu | juade<br>yikunku (alt.) | < Gã *duade*<br>? |
| | Krachi | ɔbra | ? |
| | Nkonya | gbedi | Ewe loan |
| | Larteh | agberi | Ewe loan |
| Gã-Dangme | Gã | duade | ? neologism |
| | Dangme | agbeli | Ewe loan (in recent times) |
| Gbe | Ewe | agbeli | neologism; lit., 'there is life' |
| Togo Languages | Avatime | agbeli | Ewe loan |
| | Akpafu | gbedi | Ewe loan |
| | Lefana | bàncì | Akan loan |

cassava, but instead endowed it with a name that identified its overseas origin; they constructed a neologism combining the common Akan adjective prefix, *-boro*, ('from overseas') with their name for an indigenous plant they knew as *awennade* (*Byrsocarpus coccineus*; the plant is without a 'western' name),[61] to which they thought it bore important similarity.[62] Herbalists throughout the Gold Coast valued *awennade* (literally, 'to eat iron') because they could extract the juice of the root to make an

[61] H. M. Burkill, *The Useful Plants of West Tropical Africa*, 2nd edition, (Kew: Royal Botanic Gardens, 1985–2006), vol. 1, 518.

[62] The name was recorded as *abromenade* in De Marrée, *Reizen op en beschrijving van de Goudkust van Guinea*, vol. 1, 82. This name is no longer in use; modern residents now refer to cassava by the common Akan name (*bankye*) and, sometimes, names originating from Kwa languages in Côte d'Ivoire.

analgesic. Specific uses were for aches in the mouth and throat, painful measles sores, and scrapes and abrasions from accidents with farm tools, including iron hoes and cutlasses.[63]

These forest farmers planted just a little cassava on old yam plots in case of a rare failure of their favored starches: yam and plantain, which they grew on upland gardens; and rice which they grew along the river valleys. Forest farmers who did plant cassava often intercropped it with young vegetables and new banana plants on old yam gardens, so that the quick-growing cassava could provide them with light shade. After two years, they harvested the cassava (and processed it into food, if necessary, or otherwise perhaps sold or discarded the tubers) and brought the plants underneath into full sun. Farmers who left cassava to grow on old garden plots so as to nurture fallow regeneration could use the tubers of the mature plants as an extra source of food, thus reducing the demands for (male) labor to clear new gardens annually from dense vegetative cover. However, when farmers came to reclaim these gardens from fallow and again plant yams and plantains, they found that these gardens required ever more (women's) work to weed, because the soils abounded with cassava seeds produced during the fallow period as abandoned cassava plants reached sexual maturity. Once the cover was cleared, the seeds took advantage of the sunlight and soil disturbance to sprout and become what one Dutch observer recalled as a most pernicious weed:

> [Cassava] is a sort of potato, of a peculiar type. They grow like potatoes, in mounds, with the fibrous roots of a vegetable, [and] which sends up a finger-thick shoot upon which are two or three oval leaves. Where they have been planted before, it requires a lot of strength to uproot them, and so they are a dangerous weed in the garden. When eaten, they are neither of pleasant taste nor easy to digest; [are, in taste] soapy and similar to frozen potatoes [...][64]

The shape and number of leaves described on these young plants is indicative of seedlings rather than plants grown from stem cuttings. The latter would immediately develop palmate clusters of five or more leaves, identical to that of its parent plant.

---

63 Abbiw, *Useful Plants of Ghana*, 149, 151, 173, 186. The plant is now the subject of medical research for its sedative properties; see Abidemi J. Akindele, Erika A. Eksioglu, Jason C. Kwan, Olufunmilayo O. Adeyemi, Chen Liu, Hendrik Luesch, and Margaret O. James, "Biological Effects of *Byrsocarpus coccineus in vitro*," *Pharmaceutical Biology* 49, 2 (February 2011), 152–160.

64 De Marrée, *Reizen op en beschrijving van de Goudkust van Guinea*, vol. 1, 82.

*Outsiders and Renewed Innovation with Cassava*

Slaves, bondsmen, and debtors recruited to newly established plantations in the southeastern Gold Coast after 1788 bore the burden of expanded cassava cultivation and consumption. Copenhagen-based merchants established the first plantations on the Gold Coast where substantial amounts of cassava were planted and eaten. Danish moralistic upset with the horrors of the slave trade stimulated a national ban on such trafficking in 1803 and the Crown thereafter directed metropolitan investors, and subsidized many, to establish plantations near the Guinea headquarters of Danish commercial interests in Guinea, at Christiansborg. Danish planters acquired land-rights from local notables to farm plots north of Accra, in the Legon hills and north along the Akuapem ridge, which Danish botanists had correctly identified as having the rich and well-watered soils required for intensive, plantation agriculture.[65]

Planters modeled their new enterprise on operations familiar from the West Indies, where slaves tended cash crops to be exported to Copenhagen. When the Guinea planters found hilly Akuapem terrain to be largely unsuitable for cultivating sugarcane (the most profitable of West Indian crops), they targeted indigo and cotton. Slaves bought or hired out to work these cash crops were largely left responsible for providing for their own subsistence on Danish plantations in Guinea. They bore the cost of their own reproduction during their off-hours by hunting, fishing, gathering, and tending provision gardens on marginal lands, in kitchen gardens immediately outside their simple shacks and larger gardens made on unworked areas of the plantations.[66] Cassava appears to have been a popular choice among starchy staples on such provision plots, given its low demands on soil fertility and cultivation labors. Additionally, on at least one plantation, planters organized slaves to grow cassava (as well as plantains and maize) on core plantations grounds for collective subsistence.[67]

Following the closing of the legal slave trade by the other Europeans trading at ports of the southeast, Accra-based merchants diverted some of

---

65 C. D. Adams, "Activities of the Danish Botanists in Guinea, 1783–1850," *Transactions of the Historical Society of Ghana* 3, 1 (1957), 30–46.

66 Yaw Bredwa-Mensah, "Archaeology of Slavery in West Africa" (unpublished conference paper; World Archaeological Congress 4, Cape Town, 10–14 January 1999); Kea, "Plantations and Labor in the South-East Gold Coast," 137.

67 That is, at the plantation known as 'Bibiase' in Akuapem; see Henrik Jeppesen, "Danske plantageanlæg på Guldkysten, 1788–1850," *Geografisk Tidskrift* 65, 1 (1966), 57–59.

their inventories of human capital to agricultural production of foodstuffs for local consumption. The Accra-based merchants bought land-rights in the hinterland of the city and organized their slaves to produce food to be marketed commercially to the urban, non-farming populace. On these plantations slaves grew large amounts of cassava not only for their own sustenance but also to provide raw tubers to the urban markets and consumers there who were familiar with, if not excited about, consumption of this starch.[68]

From the 1830s, two groups of Africans in the New World diaspora emigrated to the southeastern Gold Coast and invested in cassava cultivation in order to satisfy their cravings for specific cassava-based foods that reminded them of 'home.' The first group to arrive in Accra was comprised of Afro-Brazilians resettled to West Africa following a series of slave revolts in Bahia between 1831 and 1835. At least some of these immigrants were native-born Africans sent to Brazil from the Slave Coast. Some were made into slaves through violence near the coast amongst Youruba-speakers and others were war captives from the most Hausa-speaking to the north.[69] The expelled Afro-Brazilians settled in the coastal ports of West Africa where Brazilian traders had long been active, including Slave Coast. Subsequently a small number created a spur settlement on the Gold Coast at Accra.

There, Gã-speaking indigenes recognized the newcomers as a distinct community of strangers, whom they named 'Tabon,' mimicking the way the newcomers made their greeting (Portuguese, *Está bom?*, 'Is it good?'). The Tabon converted their banishment from the New World into a commercial success. They used their diasporic contacts to establish themselves as representatives of Brazilian-based merchants then selling tobacco and rum; and they used their contacts with Afro-Brazilians who had settled in leeward West African ports to receive slaves illegal to export from the Gold Coast, and transport them via overland routes to the Slave Coast, where slave-trading suppression activities were yet still easily evaded.[70]

Tabon traders invested some of their wealth in acquiring the accoutrements that, in Brazilian style, distinguished them as successful

---

68 Kea, "Plantations and Labor in the South-East Gold Coast," 125.

69 Paul Lovejoy, "Background to Rebellion: the Origins of Muslim Slaves in Bahia," *Slavery and Abolition* 14, 4 (1994), 151–80; João José Reis, *Slave Rebellion in Brazil: The Muslim Uprising of 1835 in Bahia*, rev. and expanded edition (Baltimore: Johns Hopkins University Press, 1993).

70 Richard Rathbone, "The Gold Coast, the Closing of the Atlantic Slave Trade, and Africans of the Diaspora," in Palmié, ed., *Slave Cultures and the Cultures of Slavery*, 55–66.

merchants and planters. They had substantial houses built of stone, which contrasted with local waddle-and-daub residences. They invested in land for farms. And, perhaps most significantly, they owned slaves.[71] Tabon slave-owners used some of their bondsmen to cultivate cassava on the plantations they owned and then to process the raw tuber into familiar Brazilian foods, including the toasted cassava meal which they knew as *farinha da mandioca* (Portuguese, 'cassava flour,' and most often simply referred to as *farinha*). Tabon taste for *farinha* required that they introduce new, labor-intensive processing techniques for cassava which were previously unknown on the Gold Coast.

Slaves, wives, and daughters, grated fresh cassava tubers by hand on a perforated metal sheet; then they packed the pulpy mash into a woven sieve to which they applied physical pressure, which served to press out the roots' juice (which contained most of the poison indigenous to American varieties of cassava). Because Gold Coast varieties of cassava were non-toxic, this step was unnecessary to render the local food safe to consume, but processors nonetheless followed this procedure in order to produce the semi-dry pulp required for the next stage of processing. Cooks then toasted the grated pulp over a fire on a broad iron pan to rid it of residual moisture and which left the shrunken 'gains' of pulp with a desirable toasted flavor and polished appearance. Thus they produced an ingredient that, although it was not as fine as well-ground wheat *farinha*, could be stored indefinitely, so long as it was kept from moisture or pests. These processing techniques were Amerindian in origin and borrowed by the Portuguese and their slaves in Brazil, as well as throughout the Caribbean, and in Portuguese-influenced African territories of Angola and the Atlantic islands.

Indigenous residents of West African ports probably came into contact with *farinha* first at markets where Afro-Brazilian women sold their exotic foods.[72] Local slaves who manufactured this food and the locals who

---

[71] Lisa A. Lindsay, "'To Return to the Bosom of their Fatherland': Brazilian Immigrants in Nineteenth-Century Lagos," *Slavery and Abolition* 15, 1 (1994), 22–50. Although there is a great deal of literature on the Afro-Brazilians of West Africa, the community at Accra remains understudied. The 2005 state visit by the Brazilian prime minister to Ghana and the potential for local investment by South America's fastest growing economy has generated some limited new interest. See Marco Aurelio Schaumloeffel, *Tabom: The Afro-Brazilian Community in Ghana* (self-published, 2009); Alcione M. Amos and Ebenezer Ayesu, "'I am Brazilian': History of the Tabon, Afro-Brazilians in Accra, Ghana," *Transactions of the Historical Society of Ghana*, new series, 6 (2002), 35–58.

[72] Lindsay, "'To Return to the Bosom of their Fatherland,'" 29.

purchased the meal named it *gari*, perhaps an abbreviated reinterpretation of the Brazilian name, *farinha*.[73] West Africans prized *gari* because it offered, for the first time, a technique to store cassava in a stable, ready-to-eat form. West African women applied their own creativity to this sort of cassava processing and developed a way to bypass the tough physical work of pressing a sieve full of pulp by hand. Instead they built mechanical presses which used stone weights and gravity to express the juice; although these presses required more time than a hand-press to fully drain away the juice, this method allowed airborne agents to inoculate the mash and induce fermentation, which lent the final product the pungent, sweet-sour taste local consumers preferred.[74]

In the 1840s a second, smaller wave of New World immigrants arrived on the Gold Coast. These were Jamaicans, who introduced a way of eating cassava born of Amerindian tradition but which bore a name from the Niger Delta. In the eighteenth century, Ngwa-Igbo-speaking slaves brought to the British West Indies applied a name they knew, *gbam-gbam*, meaning 'a plate for eating,' to the hard, flat, round cakes of grated cassava that were to become the starchy staple of their slave diets. Perhaps it was the shape of the cakes that encouraged this naming. Or perhaps it was the function of the cake to hold other, more flavorful accompaniments—most often protein-rich 'soups' that were, and continue to be, characteristic of West African and Afro-Caribbean cuisine.[75]

These cakes were in fact a borrowing from indigenous Amerindian foodways; Arawak-speaking Amerindians named these cakes *cassaba*, a

73 *Gari* is a name now part of most coastal West African languages. Elsewhere it has been suggested that *gari* was derived from the Hausa *gari* ('any flour') and was introduced via Yoruba (*gari*) to Gold Coast languages (cf. Ewe, *gali*); M. E. Kropp Dakubu, "A Survey of Borrowed Words in Dangme," *Research Review*, supplement no. 4 (1973), 116. Two interpretive points cause me to disagree. First, all the coastal languages of West Africa, however, use the name specifically for fermented cassava meal. Second, I think mere similarity in the term fails as an historical explanation when there is no other evidence of any other contemporary contact between the communities—and there is indeed none, at least until early twentieth century when British colonizers recruited in Northern Nigeria and placed men (who were presumably conversant in Hausa) in the Accra constabulary. There is yet another socio-linguistic alternative: that these immigrants *while still in Brazil* had used the Hausa word for 'millet flour' to denote what others there called *farinha de mandioca* (made from cassava, but unfermented) and they in turn brought this word and the culinary form to Accra.

74 In modern Brazil, cassava processors use similar techniques to make the fermented *farinha* known as *polvilho*.

75 Richard Allsopp and Jeannette Allsopp, *Dictionary of Caribbean English Usage* (Oxford: Oxford University Press, 1996), 76.

form of which was first noted in the journal of Christopher Columbus ('...their bread which they call *caçabi*').[76] Amerindians prepared the cakes by grating the tubers; then ridding the grated mush of its poisonous juice through pressing and washing; then pressing this starchy matter inside a metal collar to give it form; and finally cooking them on a hot pan (with or without oil). Amerindians favored *cassaba* because, when cooked hard and until dry, it stored well—like Europeans' hardtack. Slave-owners favored cultivation of cassava plants because they furnished much bulk starch even when grown on nutrient-exhausted or otherwise poor ground, which allowed prime plots to be devoted to more lucrative crops.

West Indian slaves found ingenious ways to improve the flavor and texture of the cakes and to otherwise introduce variety to their diets by adding spices, grated coconut, cut vegetables, and other flavorings. Enslaved Afro-Jamaicans knew these cakes as *bammies*, while elsewhere in the British Caribbean slaves knew other similar names for this food (e.g., *bambam* on Greneda and St. Vincent; *bambula* on Antigua).[77] Slaves in Brazil, and the Spanish and French colonies, appear to have borrowed directly from the Amerindian word for these cakes, and knew (and know) them as *cassava* (or *casaba, cazaba*). The Brazilian Portuguese term for the cake, *cassava*, has been borrowed in West Africa by Yoruba-speakers to name the plant (*kasada*, showing a typical /v/ to /d/ shift, owing to the lack of the /v/ sound in Yoruba).

The Jamaican immigrants were recruited from the island's Moravian church community by the Basel Missionary Society, which was then active in the southeastern Gold Coast. The Jamaicans' sponsors arranged for them to take up residence at the mission station at Akropong, where they were expected to pursue a civilizing campaign comprised of Christian conversion and commercial agriculture. Although the immigrants failed to make much of a religious impact, they did achieve a culinary one by introducing American foodways they knew. The Jamaican immigrants introduced

[76] Cecil Jane, ed., *The Journal of Christopher Columbus* (New York: Bramball House, 1960), 125.

[77] Allsopp and Allsopp, *Dictionary of Caribbean English Usage*, 76–77. The paucity of sources for early Caribbean creoles makes it difficult to untangle the linguistic history of the term. Although New World variants of the Ngwa-Igbo *gbam-gbam* appear in modern speech of many islands of the formerly British West Indies, these might be the result of more modern labor migrations (e.g., Jamaican migrants might have introduced the term to Belize Kriol during heavy labor migration there in the early 20th-century expansion of Central American commercial banana farming).

West Indian cocoyams, breadfruit, mangoes,[78] and the ackee fruit.[79] The Jamaicans also cultivated cassava—perhaps with the help of slaves they eagerly accumulated, much to the distress of the mission elders—and processed these tubers into *bammy* cakes.

By the late 1850s, people beyond just the Jamaican community at Akropong were consuming *bammy* cakes, and its name was common among Gã-speakers of Accra.[80] These Jamaican foodways largely faded with the passage of time and the assimilation of the expatriates into local communities, and perhaps only two relics of their culinary tradition survive in Ghana today: the first is the Gã name *bami* for a specific cassava cultivar;[81] and the second is the Ghanaian dish now known as *gari fortor* which, like the native Jamaican elaboration on *bammy*, mixes local flavorings (onions, tomatoes, eggs) into the shredded cassava before frying.

### *Colonial Postscript*

By the end of the nineteenth century, farmers throughout all regions of the Gold Coast included some cassava cultivation in their food strategies, as a supplement to the yams, plantains, and grains they preferred. While a small number of immigrants and their descendants maintained fond, foreign notions of specific cassava foods, most Gold Coast consumers recognized that eating cassava indicated social degradation arising from enslavement, impoverishment, and general misfortune. In the forests and savannas west of the Volta River, those who could afford to eat other starchy staples always did.

Widespread adoption of cassava as a *principal* starch trailed the expansion of commercial cocoa (another foreign food crop, introduced to the Gold Coast from the Americas, via Fernando Po, in 1878) farming in the twentieth century. This rapidly expanding commercial farming sector drew many poor, migrant laborers (and hired-out slaves), who grew cassava because they could produce much raw bulk starch to be eaten in

---

[78] Reindorf, *History of the Gold Coast and Asante*, 269.

[79] Roger Blench, "Unnatural Terms for Natural Things: Comparative Ethnobotany in the Volta Region of Ghana and Its Implications for Historical Linguistics" (unpublished conference paper; Colloquium of African Languages and Linguistics, August 2000, Leiden).

[80] Rev. Johannes Zimmermann, *A Grammatical Sketch of the Akra or Ga language ... a vocabulary of the same, with an appendix on the Adanme dialect* (Stuttgart, 1858).

[81] M. E. Kropp Dakubu, *Ga-English Dictionary* (Legon: Institute of African Studies, University of Ghana, 1973); here *bami* is listed among some fifty other specific cultivars known to Gã-speakers.

simple ways (*fufu, ampesi*) with little effort and without sacrifice of prime garden environments, which they reserved for cocoa. Cassava cultivation probably reached historical peak importance for most Gold Coast farmers and consumers during a brief period in the 1920s and 1930s. The flu pandemic of 1918–19 left Gold Coast farmers physically weakened and short of manpower, and they therefore turned increasingly to cassava because they could rely on this drought- and disease-resistant crop to produce the most starch for the least labors.[82] But Gold Coast farmers lost their unreserved confidence in reliable cassava bounty increasingly after 1926, when their plants began to change leaf color, then withered, and eventually died from a previously unknown disease. By 1936, colonial crop scientists had identified the disease as a fly-borne virus named cassava mosaic, but by then the disease had become uncontrollable, endemic, widespread, and devastating. Only with scientific development of moderately disease-resistant strains and topical insecticides after the Second World War did Gold Coast farmers begin to once again consider cassava an important starch,[83] but never again did cassava return to the position it had held during the brief 1918–1936 window as the principal starch in their diets.

The exceptionality of this brief heyday of cassava cultivation and consumption was missed by colonial agricultural officials who described the colonial moment in the food economy as 'traditional.' Historians who borrowed from their reports portrayed cassava anachronistically as the most significant starch—in terms of bulk and caloric contribution—in Gold Coast diets since early Atlantic times.[84] Such ahistorical perspectives also miss the uniqueness of Gold Coast dynamic experimentation with and adaptation of cassava as they integrated themselves into the commercialized slaving of the pan-Atlantic world.

## *Conclusion*

Gold Coast farmers, unlike any other agricultural community in the Atlantic world, made cassava their own on their own. After the initial European introduction of the raw genetic materials for experimentation

[82] Here I am extrapolating from elsewhere in West Africa; see D. C. Ohadike, "The Influenza Pandemic of 1918–1919 and the Spread of Cassava Cultivation on the Lower Niger: A Study in Historical Linkages," *Journal of African History* 22, 3 (1981), 379–92.

[83] Doku, *Cassava in Ghana*, 19.

[84] Wilks, *Forests of Gold*, 40–90.

within the rudimentary local slave economy, Gold Coast farmers applied their own ingenuity to develop cassava as a food crop for themselves, without any tutelage by foreigners who knew how to plant and process the poisonous plant. Isolation from such agro-technological and culinary traditions favored the development of an indigenous cassava culture characterized by their creation of non-toxic cultivars and positioning cassava in their food strategies not as a staple starch but rather as a supplementary insurance against crop failure, hunger, and debts which might make them vulnerable to pawning and enslavement.

CHAPTER SIX

# FINDING FOOD IN AFRO-ATLANTIC HISTORY

Foodways are fundamental expressions of human agency and solidarity. Multiple material and cognitive contexts give meaning to human experience and frame possibilities for action. Fastidious attention to motivated strategizing and the distinctive epistemological elements of History as a way of knowing and understanding—context, change, agency—reframes the African narrative of the 'Columbian Exchange' away from the 'spread' of crops and the passivity that such a construction imposes on the people who chose them, ate them, bred them, and rebuilt their lives around them. When sparked by doubt about what lies beneath such opaque abstractions, asking the basic questions of historical research—*who*? did what? when? how? where? and why?—generates new lines of inquiry which, with appropriate disciplinary agility, produce richer, more complex, and, most significantly, recognizably *humanistic* histories.[1]

Cultivating, cooking, and breeding are altogether appropriate to explaining how people engaged the qualities of the cultivated starchy crops, none of which had the physical means to 'spread' on its own without human actions. Ears of maize and Asian rice held back their seed from falling to the ground, plantains and bananas produced new shoots which grew up immediately and unpromisingly close to the parent plant, and cassava had no fruits to attract animals that would deposit their seedy contents further afield. These botanical barriers to geographical expansion independent of human actions are important not only when describing the transoceanic adoption of American crops in West Africa but also the crops' subsequent areal expanse. Retelling the history of the Atlantic-era Gold Coast in terms of African cultivators and cooks—their perceptions and their actions—rather than in terms of such abstractions as the 'spread' of

[1] Jan Vansina, "The Power of Systematic Doubt in Historical Enquiry," *History in Africa* 1 (1974), 109–27; David Henige, "Gambit Declined: Pervasive Doubt About Systematic Doubt," in Robert W. Harms, Joseph C. Miller, David C. Newbury, and Michele D. Wagner, eds., *Paths Toward the Past: African Historical Essays in Honor of Jan Vansina* (Atlanta: African Studies Association, 1994), 77–95; Joseph C. Miller, *The Problem of Slavery as History: A Global Approach* (New Haven: Yale University Press, 2012).

crops or the 'impact' of overseas slave trade or the Europeans who visited these shores, or the institutional histories of the foreign companies that operated there moves the narrative towards African agents and domestic themes and rhythms of change. West Africans drew on the possibilities—including genetic resources, cultural knowledge—that outsiders (and, to be emphasized, not just Europeans) brought to their shores in order to fashion their own ever-changing strategies for physical sustenance, hopes of economic gain, and lives made meaningful by culturally and historically specific ideas about what they ate together. The agricultural and cultural changes they pursued show multiple streams of historical interaction in Atlantic history, not all of which run from 'Old World' resources toward 'New World' developments. Nor can they be adequately explained within solely continental African contexts.

Explaining historical change in West Africa requires understanding events, ideas, and processes far afield.[2] Hence this project has Iberian hunger stimulating Portuguese experimentation with food production on the Gold Coast, including the transportation and sponsorship of trial gardens in the contexts of these foreigners' notions and anxieties about eating and health. It then highlights the seeds, tastes, and insights of enslaved Africans from distant parts of the African littoral as they integrated as foreigners into local communities in West Africa. Then it follows some of these captives across the Atlantic as they continually reinvented new cuisines and names. And then their American creations flowed back to Africa, as people from the diaspora—now foreign Afro-Jamaican and Afro-Brazilian heirs to their African ancestors' culinary creativity—brought their own exotic foodways to the palette of Gold Coast choices. West Africans engaged the nascent Atlantic world not with passivity, but rather with creativity as full partners with those on neighboring continents whose histories have enjoyed longer, and greater, scholarly attention, to create these fusion foodways.

[2] Therefore '*cis*-Atlantic' in the tripartite typology of David Armitage, "Three Concepts of Atlantic History," in David Armitage and Michael Braddick, eds., *The British Atlantic World, 1500–1800* (New York: Palgrave Macmillan, 2002), 11–30.

## *Reflections & Prospects*

*The most urgent and immediate challenge is to restore Africa to the Atlantic.*
Allison Games (2006)[3]

*If the category of the Atlantic is to mean anything, it ought to include Africa, but there seems to be no room for this often overlooked fourth continent in most new versions of the Atlantic.*
Jorge Cañizares-Esguerra (2007)[4]

Fortunately, Africa's absence is not so dire nor the situation terminal. Because the condition, to the extent it does exist, is in fact largely self-imposed by the Europe-centered imperial framing of Atlantic history that scholars have drawn on, the remedies are in their very own hands. They frame their understandings in phrasings that signal their intellectual orientations to their readers. 'Atlantic,' with regard to Africans' foodways along the Gold Coast, is such a word, invested with meanings encompassing various temporal, geographic, and intellectual subjectivities. As a field of scholarly endeavor, the 'Atlantic' has grown in the last two decades with remarkable vigor, putting out new shoots from deep roots in the isolated units of 'nations' that have defined History since its birth as a modern academic discipline in nineteenth-century Germany.[5] Early efforts toward what became 'Atlantic history' began with European historians of empires who valorized the achievements of the monarchies

---

[3] Allison Games, "Atlantic History: Definitions, Challenges, and Opportunities," *American Historical Review* 111, 3 (2006), 754.

[4] Jorge Cañizares-Esguerra, "Entangled Histories: Borderland Historiographies in New Clothes?", *American Historical Review* 112, 3 (2007), 787–99.

[5] Discussions include, but are not limited to, papers from "Round Table Conference: The Nature of Atlantic History" (1998), published as proceedings in *Itinerario* 23, 2 (1999); Nicholas Canny, "Writing Atlantic History, of Reconfiguring the History of Colonial British America," *Journal of American History* 86, 3 (1999), 1093–1114; William O'Reilly, "Genealogies of Atlantic History," *Atlantic Studies* 1 (2004), 66–84; Bernard Bailyn, *Atlantic History: Concept and Contours* (Cambridge MA: Harvard University Press, 2005); AHR Conversation: "On Transnational History," *American Historical Review* 111, 5 (2006), 1441–64; Jorge Cañizares-Esguerra, and Erik Seeman, eds., *The Atlantic in Global History: 1500–2000* (New York: Prentice Hall, 2007); Alison Games and Adam Rothman, eds. *Major Problems in Atlantic History* (Boston: Houghton Mifflin, 2008). The Africanist perspective is from Joseph C. Miller, "History in Africa / Africa and History," *American Historical Review* 104, 1 (February 1999), 1–32; Tobias Greene, "Building Creole Identity in the African Atlantic," *History in Africa* 36 (2009), 103–25; James H. Sweet, *Domingos Álvares, African Healing, and the Intellectual History of the Atlantic World* (Chapel Hill: University of North Carolina Press, 2011), 4–8.

behind modern nations as they understood them: planting flags, creating colonies, and bringing civilization. In this case of food, they credited European adventurers and scientists with 'spreading' new crops of supposed superiority by using the technologies that only they had, including ocean-going ships fitted with sails, cartography, and the burgeoning field of scientific botany. This triumphalist narrative was very masculine and looked at the Atlantic as a periphery of metropolitan centers.[6] This scholarship faithfully reproduced the historical sources themselves, which were European, documentary, and loyal.

Scholars of early West African history helped spark the very idea of 'Atlantic History' in the 1960s during the heyday of what was then known as 'pre-colonial' history.[7] Seeking sources that were (or so at least they initially thought) untainted by European influence, they engaged non-documentary sources based in Africa—particularly oral histories, languages, and physical debris—and they borrowed existing methodologies from sibling disciplines to convert these local materials into historical data for progressive narratives meant to show a record of African political and economic achievement before European colonialism interrupted this progress toward a European-like modernity. Africanist historians (as well as Afro-Americanists) worked against some skepticism about their sources and methods from professional peers who studied American and European history as articulated in the expository documents they regarded as valid sources, and amid polarizing racial politics.[8]

Conveniently, the African oral traditions notably highlighted seemingly recognizable kings and royal courts and thus had very much to offer for scholars of political history. So rich seemed the yield that no one noticed the absence of attention to food. Archaeology captured a rather broader swathe of past African societies, though its yield was restricted in its coverage for rather practical reasons: it was difficult to find sites left behind by low-impact inhabitants of the forests and moist savannas.[9] Also, the limited material culture that survived to be discovered by then-existing

---

[6] For a feminist contrast, see Londa Schiebinger, *Plants and Empire: Colonial Bioprospecting in the Atlantic World* (Cambridge MA: Harvard University Press, 2004).

[7] Games, "Atlantic History"; Miller, "History in Africa / Africa and History."

[8] For some personal reflections on the juxtapositions with United States politics, see Jan Vansina, *Living with Africa* (Madison: University of Wisconsin Press, 1994); Richard Price, "The Miracle of Creolization: A Retrospective," *New West Indian Guide* 75 (2001), 35–64. Also Richard Reid, "Past and Presentism: The 'Precolonial' and the Foreshortening of African History," *Journal of African History* 52, 1 (2011), 135–55.

[9] Above; chapter 2, note 20.

recovery and dating methods showed few long-term processes of change—almost to the point of 'continuity'—that were difficult to reconcile with the 'events' recalled in oral histories as 'migrations' or magico-cultural interventions.[10] Oral history came to be appreciated by outsiders as better understood as historiography, that is as subjective interpretations of the past that skillful performers tailored to the immediate circumstances of their audiences.[11] Very little of the content of the oral traditions elicited by historians of politics concerned either farming or food.

A very few historians used historical linguistic methodologies to reach into the deep African past. Unfortunately, this work concentrated on the numerous and relatively clearly related Bantu languages spoken in the Central African forests and in Africa's Eastern and Southern Savannas rather than on the tongues of West Africa. The field of historical linguistics in West Africa remains underdeveloped compared to other parts of the continent, and detailed descriptions of phonology and vocabulary for the historic forms of languages of West Africa's Atlantic-era past do not exist.[12] West Africa's languages feature less morphology—which is a linguist's way of saying that the languages contain fewer vowel and consonant sounds and that the words are shorter. There are therefore fewer sound features for scholars to analyze, compared to the Bantu languages spoken throughout Central, East, and Southern Africa. The lexicostatistical comparisons of modern words used to reconstruct putative ancestral words and to classify languages in terms of shared core vocabulary is frustrated in West Africa by the ambiguity of the ultimate data. Some students of

[10] E.g., the spirit 'washing' the toxins from cassava [above; chapter 5]. Oral traditions from the both colonial and post-colonial era point to 'migrations' at standard generational-length-reckoned time-depths of approximately five hundred years ago; hence the date of the arrival of distant ancestors claimed by modern communities ('the Fante,' 'the Ewe,' *et al.*) is something of a moving target for historians who use these to pin calendrical dates, and has moved from around 1400 CE to around 1500 CE. For nineteenth-century traditions of prior migrations, see Reindorf, *History*, 44; Bowdich, *Mission from Cape Coast Castle to Ashantee*, 229. For academic interrogation see, Wilks, "Land, Labor, Capital." The oral historians' stress of 'migration' to explain change is counterbalanced without any sense of contradiction by explanations of extreme indigeneity; e.g., the Asante explanation that ancestors came out of holes in the earth around Kumase.

[11] For Africa, see Henige, *Oral Historiography*; Miller, *The African Past Speaks*; Thomas Hale, *Griots and Griottes: Masters of Word and Music* (Bloomington: Indiana University Press, 1998).

[12] I do not think there has ever been a good explanation of this phenomenon in print, so I can only give my sense as an outsider to the field. The best presentation of the history and prospects for the field is Gerrit J. Dimmendaal, *Historical Linguistics and the Comparative Study of African Languages* (Amsterdam and Philadelphia: John Benjamins, 2011).

West African tongues have turned to classifying languages by their grammatical systems, including techniques known as syntactic and qualitative typology. Because these are qualitative rather than quantitative methodologies, there is a considerable array of differing sketches of the possible genetic degrees of relatedness between languages.[13] Thus, compared to the histories of the Central African peoples who engaged the possibilities of Atlantic trade, historically minded linguists and historians have only recently begun to develop the potential of the languages to reveal historical change in of the lives of early speakers in West Africa.[14]

In partial compensation, a much richer and longer-sustained effort was directed at the principal European travelogues that described West African societies in the Atlantic era. Initial uses of these seemingly straightforward sources to find confirming convergences seemed to some critics to verge on abuse of scholarly rigor.[15] In regards to farming and food, potential errors crept into the narrative through translator-editors' eagerness to cross-reference evidence of crops and cuisine from other critical editions, which was understandable, while the topical field of early African history fell fallow. Critical clarification has been stubbornly pursued, and notably by European scholars, whether by language training or professional climates that regarded editorial precision as real scholarship. The documents available now approach a critical mass that enables historians to engage major questions of relevance to Atlantic history—including the historical development of slavery as a local institution—that were raised more speculatively a generation ago.[16] As historians in the 1990s and after shifted their attention to colonial and post-colonial histories of Africa,[17] it was historical archaeologists who undertook the most original research in the pre-colonial past.[18] They drew on the critical editions to contextualize

[13] There is a new attempt to add quantitative reasoning to checklist grammatical features; see Angela Kluge, "Qualitative and Quantitative Analysis of Grammatical Features Elicited among the Gbe Language Varieties of West Africa," *Journal of African Languages and Linguistics* 27, 1 (2006), 53–86.

[14] Fields, *Deep Roots*.

[15] Heintze and Jones, eds. "European Sources for Sub-Saharan Africa before 1900: Use and Abuse."

[16] Bruce Mouser, "A History of the Rio Pongo: Time for a New Appraisal," *History in Africa* 37 (2010), 329–54.

[17] Jan Vansina, "The Doom of Early African History?," *History in Africa* 24 (1997), 337–43; Reid, "Past and Presentism"; Miller, "History in Africa / Africa in History."

[18] Vansina, "Historians, Are Archaeologists Your Siblings?"; Robertshaw, "Sibling Rivalry?"

the remains they found, though not always with critical understandings of the texts.[19]

## *Fusion Foodways*

This social history has focused on changes in the elemental aspects of the everyday lives of the majority of the people who farmed and, of course, of everybody who ate. It highlights the prominent role of women, who in the Atlantic world served as 'gatekeepers' to the introduction of new crops into domestic gardens and then into their kitchens.[20] They devised ingenious ways of cooking frugal meals. They developed innovative culinary repertoires to moderate the monotonous reliance on small sets of locally dominant starchy staples. Variety offered modest nutritional benefit, but cooks were more likely inspired by their own creative urges and appreciation of the aesthetics of feeding. Focusing on their techniques of food preparation and the culinary forms they produced gives historians a glimpse into domestic realms in Atlantic-era West Africa that are otherwise elusive. This gaping hole in the quality of lives certainly stems from the obscurity of women's private lives in the most commonly used documentary sources. The social history of foods and feeding may also have been considered mundane for the first scholars—mostly male anthropologists—who studied African foods exclusively for their symbolic significance as totems and taboos.[21] Early (and even colonial) Africa also lacks the cookbooks that have provided scholars researching other historical food cultures with clear indications of common foods and their preparations, tastes and manners.[22] However, it is possible to reconstruct much of Gold Coast women's daily work, and their perception of it, from names for specific foods mentioned in contemporary sources along with scattered evidence of their preparation. And evidence from the Atlantic world indicates that women valued their work with foodways as a means to express their own aesthetic freedoms. In doing so, they creatively developed new

---

[19] Jones, "Drink Deep, Taste Not."

[20] Wm. Alex McIntosh and Mary Zey, "Women as Gatekeepers of Food Consumption: A Sociological Critique," *Food and Foodways* 3, 4 (1989), 317–21.

[21] Sidney Mintz, *Sweetness and Power: The Place of Sugar in Modern History* (New York: Penguin Books, 1985), 3.

[22] For an excellent example, see Jeffrey M. Pilcher, *¡Que vivan los tamales!: Food and the Making of Mexican Identity* (Albuquerque: University of New Mexico Press, 1998). For post-colonial recipes from Ghana, see Fran Osseo-Asare, "We Eat First with Our Eyes: on Ghanaian Cuisine," *Gastronomica*, 2, 1 (2002), 49–57.

communities of belonging through sharing and eating meals prepared with ingredients and techniques that referenced their knowledge of cross-currents of history.[23]

For so long as the history of farming and food in West Africa remained a neglected field of inquiry, historians expressed less than the full complexity of their African agents of American change. Certainly arguments about the talents of Africans forced into American fields need to be grounded in knowledge of not only the specifics of *whence* the slaves came (and this aspect is difficult to deduce even from the best knowledge base of the trans-Atlantic trade,[24] where points of departure from the coast are clear but not the slaves' original geographic—and therefore ethnic/cultural provenances)[25]—but also *when* they departed from a continent undertaking great reorganizations of communities and significant changes of core crops.

This study also explores the cognitive contexts that motivated cooks and cultivators to adopt and adapt the cultigens that they did. Though eating serves a fundamental biological purpose, they did not decide to favor certain things to eat among a broad palette of choices did solely by rational comparisons of the nutritional benefit of the various options, as described by modern science, in pursuit of evolutionary advantage.[26] Rather, choices of what to eat are fundamental expressions of human

---

[23] Stephan Palmié, "Ackee and Saltfish vs. Amalá con Quimbombó?: a Note on Sidney Mintz' Contribution to the Historical Anthropology of African America Cultures," *Journal de le société des américanistes* 91–92 (2005), 89–122; Stephan Palmié, "Intangible Cultural Property, Semiotic Ideology, and the Vagaries of Ethnoculinary Recognition." *African Arts* 42, 4 (Winter 2009), 54–61. The literature for gender and cooking is better developed for the African diaspora in the New World, than for Africans displaced within the continent; see Robert Hall, "Savoring Africa in the New World," in Herman J. Viola and Carolyn Margolis, eds., *Seeds of Change: A Quincentennial Commemoration* (Washington DC: Smithsonian Institute Press, 1991), 161–77; Anne Yentsch, "Hot, Nourishing, and Culturally Potent: The Transfer of West African Cooking Traditions to the Chesapeake," *Sage: A Scholarly Journal on Black Women* 9, 2 (1995), 15–29; Josephine A. Beoku-Betts, " 'We Got Our Way of Cooking Things': Women, Food, and Preservation of Cultural Identity among the Gullah," *Gender & Society* 9, 5 (1995), 535–55; Carney and Rosomoff, *In the Shadow of Slavery*.

[24] David Eltis, Stephen D. Behrendt, David Richardson, and Herbert S. Klein, eds., *The Trans-Atlantic Slave Trade Database* <www.slavevoyages.org>.

[25] Eltis, Morgan and Richardson, "Agency and Diaspora in Atlantic History."

[26] Such as done by the 'cultural materialists' of food studies, who describe subsistence strategies as designed to reap optimal nutrition within the constraints of the physical and economic (labor) environment. See, for example, M. Harris, "Foodways: Historical Overview and Theoretical Prolegomenon", in M. Harris and E. B. Ross, eds., *Food and Evolution: Toward a Theory of Human Food Habits* (Philadelphia: Temple University Press, 1997), 57–90.

feelings engaged with multiple material and cognitive contexts. Scholars have typically created an arbitrary divide between the material context (including the physical environment, technology, and social relations) of agricultural production and the cultural aspects of cooking and eating food. But farming methods and cooking practices, for their technological qualities, are no less 'cultural' in their felt meanings; those who engage technologies understand—or better, experience—them in specific cultural terms.[27] And because farming chores, cooking, and eating are activities done on a regular (daily) basis, such quotidian matters may have held more resononance for them than did seasonal harvest festivals or food taboos which were conspicuous enough to capture the attention of ethnographers.

Words—gleaned from early documents but also from modern languages—offer an opportunity to move beyond the limitations of documents written by outsiders. They are what people cared enough about to articulate. They may be used to deceive, but their existence is unwitting evidence of experiences. The formal linguistic classification of local vernaculars matters little to revealing Gold Coast foodways in the Atlantic Era, because the project does not reconstruct hypothetical root words or depend on glottochronology to estimate time depths as the Bantuists do in the early eras of the areas to the east and south. Instead the method here focuses on borrowed words and neologisms to illuminate Atlantic-era contacts (in the form of loan words and loan-translations) and *mentalités* (from the semantics of neologisms). These lingustic sources and insights from language are used in both African history and in Afro-American history—in the latter case, particularly for Maroon communities.[28]

Some of the most celebrated work in Afro-Atlantic history in recent years has targeted Africans' agricultural contributions to the development of American plantations. But this literature is dominated by Senegambia and the Upper Guinea coast, the history of rice, and the themes of inheritance and transmission of African traditions. *Fusion Foodways* adds new places and new crops to this rich scholarly debate. As an intellectual

[27] Peter R. Schmidt, ed., *The Culture and Technology of African Iron Production* (Gainesville: University Press of Florida, 1996).

[28] For Africa, see McCaskie, *State and Society*; Schoenbrun, *A Green Place, A Good Place*; Jan Vansina, *Paths in the Rainforest* (1988) and *How Societies are Born* (2004), Chris Ehret [various 1971–2011], and others. For maroons of Suriname, see Richard Price, *First Time*; and Richard Price, *Travels with Tooy*, 308–89 ["Coda: Esoteric Language—Tip of the Iceberg"]. For Jamaica, see Kenneth Bilby, *True-Born Maroons* (Gainesville: University Press of Florida, 2005), 475–83.

project, it does attempt to prove the worth of Africans to readers unfamiliar with Africa by showing their contributions to the colonial societies and economies of the Americas more familiar to them. Incorporating Africa into Atlantic history does not depend on the movements of Africans to the Americas. Instead this project focuses on African history on its own terms, endeavoring to make the history of Africans *in Africa* integral to understanding Atlantic history as a whole. Most Africans in the Atlantic era never left Africa. Approximately twelve million Africans were forced onto ships destined to American ports; another eight million went to North African and Mediterranean markets; and some four million were sent across the Indian Ocean. But more than 24 million enslaved persons were torn from their natal communities of belonging and moved to new situations within Africa, where they had to reinvent new lives of meaning through stitching together new networks of social interaction and obligation (often in the characteristic African idiom of 'family').[29] The numbers evidence what historians have agreed: that the Atlantic trade in captives was an aspect of slaving in Africa. Most of the captives retained were shuffled to social environments in Africa that were new to them, which left them isolated and dependent and therefore vulnerable. This 'hidden diaspora' and the history of Africans *in Africa* is essential to understanding Atlantic history.[30]

## *Conclusion*

> *Genuine research is never useless. It always brings something new, even if it rarely leads to the full gratification of expectations.*[31]

Exploring the Atlantic-era African past therefore requires complementary researches into agency and its constraints, instrumental technology and the culture that gives it meaning, modern clues and historical detours, and

[29] The general numbers and the highlight on African (domestic, continental) slave trades as the source of exports come from Larson, *History and Memory in the Age of Enslavement*, 268–88.

[30] Joseph C. Miller, "Retention, Re-Invention, and Remembering: Restoring Identities Through Enslavement in Africa and Under Slavery in Brazil," in José C. Curto and Paul E. Lovejoy, eds., *Enslaving Connections: Changing Cultures of Africa and Brazil during the Era of Slavery* (Amherst NY: Prometheus/Humanity Books, 2003), pp. 81–121; Larson, *History and Memory*, 268–88; Law and Lovejoy, "The Changing Dimension of African History: Reappropriating the Diaspora."

[31] Vansina, "The Doom of Early African History?"

between the agricultural (that is, techniques) and the agrarian (people).[32] The principal challenge to incorporating Africans in Africa into Atlantic history has been methodological and disciplinary. But they emerge as dynamic participants in a rigorously historical understanding of Africa in the Atlantic era with the aid of sibling disciplines. The resources for Atlantic African history are scarce only if scholars limit themselves to the written record. The overwhelming majority of people described in this study did not write their memoires, just as most enslaved African descendents in the Americas, and they were beneath the notice of European observers of matters of state and trade. Although they did not write anything down, they left records of what they did, accessible to the alert historian. In West Africa outsiders' written records scarcely illuminate anything but the cannon's shot inland covered by foreigners' diaries.[33] Increasingly, as the Europeans turned to slaving at the end of the seventeenth century, their writings became more numerous but turned to feed, not *food*, and provisioning rather than *people*. The paucity of documentary sources for the popular culture of these communities demands that scholars develop insights from other sources, drawing on the methodologies of other disciplines. This project makes use of artwork, ideas about taste and nutrition, taboos, and more—but especially language (e.g., proverbs and oral histories) and linguistics (loan words that point to past social contacts and etymologies of neologisms that reveal what people had in mind when new cultigens caught their attention).

The hallmark of academic endeavor in early African history has been its methodological agility, and thus it has much in common with the field of Native American history, where the question of *how* scholars might develop historical understandings is intertwined with the quest to discern and explain *what* people in the past did. All historians, whether academic, 'traditional' or popular-vernacular, interpret pasts for themselves, but their narratives are always performances directed at audiences of people living in a web of social relations.[34] Where scholarly monographs once

---

32 In his "Agriculture and African History," James McCann set the bar quite high: 'The ability to understand technical data on agronomy, soil science, and epidemiology in combination with historical empiricism using oral data, archival records, and even participant observation will be essential to historical reconstruction of ecological systems.'

33 The Portuguese colony of Angola is something of a source exception, due to reports from the inland areas during wars of the *conquista*.

34 On the performative aspect, see Price, *Alabi's World*, 285; Greg Denning, "A Poetic for Histories: Transformations that Present the Past," in Aletta Biersack, ed., *Clio in Oceania: Toward a Historical Anthropology* (Washington DC: Smithsonian Institute Press, 1991),

positioned methodological considerations in prefaces—at times taking almost a preemptive, apologetic tone for the evidence examined, and often explaining that the text that followed had been spared from conditional phrasings ('maybe,' 'perhaps') for the sake of the readers' patience—this project has embedded its interpretative methods and problems within the text.[35] Here, this supporting conversation with the reader is carried on in the footnotes, in common guild practice, and the methodologies introduced and discussed in the text itself at the moments when readers' attention is demanded.

Doing history demands that scholars make sense of scattered shards of the past left behind by people who are complete strangers—remembering the aphorism that, 'the past is a foreign country: they do things differently there.'[36] It also requires making sense of people in the past as *they* understood themselves, and what they thought they were doing and who they knew they were, not just how we now might also understand them.[37] Certainly all of the African farmers described here had no thought that they were engaged in making the Atlantic world (and remaking themselves) when they named so many new crops and other technologies as 'from overseas' and made them their own. But they were parts of its many-layered exchanges as much as other Africans enslaved in the Americas or European planters profiting from Asian sugarcane. To exclude them from 'the Atlantic' seems not only to neglect their achievements but also leave 'the Atlantic' incomplete.

---

347–80. For the popular-vernacular, see Chitralekha Zutshi, "Debating the Past: Academic and Popular Histories in India," *AHA Perspectives* 47, 9 (December 2009).

[35] For example, Miller, *Kings and Kinsmen*, x–xi: 'Given the limitations created by the [source] conditions just outlined, I have attempted to steer a course between the Scylla of acknowledging the imperfections of the data to the extent of saying nothing at all and the Charybdis of striving for a coherent reconstruction of events that pushes the data beyond their inherent limitations.'

[36] L. P. Hartley, *The Go-Between* (various editions, but originally published in London, 1953), first sentence of the prologue.

[37] Stephan Palmié, "Creolization and Its Discontents." *Annual Review of Anthropology* 35 (2006): 433–56. Revealing 'reverse' narrative tactics have been exploited on either side of the Atlantic for 'native' history. See David Northrup, *Africa's Discovery of Europe, 1450–1850* (Oxford: Oxford University Press, 2002); and Daniel K. Richter, *Facing East from Indian Country: a Narrative History of Early America* (Cambridge MA and London: Harvard University Press, 2001).

# WORKS CITED

## I. *Archival Sources*

Note: the following refer to collections; full references to particular items are given in the footnotes.

BNA
the National Archives: Kew, the United Kingdom.
T70/– Treasury Papers, African Companies

KITLV
het Koninklijk Instituut voor Taal-, Land-, en Volkenkunde: Leiden, the Netherlands.
Bijzondere Collecties

NA
het Nationaal Archief: The Hague, the Netherlands.
NBKG Nederlandsche Bezittingen ter Kuste van Guinea (access #1.05.14)
OWIC (Oude) West-Indische Compagnie (access #1.05.01.01)
TWIC (Tweede) West-Indische Compagnie (access #1.05.01.02)

## II. *Unpublished Works*

Blench, Roger. "Unnatural Terms for Natural Things: Comparative Ethnobotany in the Volta Region of Ghana and Its Implications for Historical Linguistics." Unpublished conference paper; Colloquium of African Languages and Linguistics, August 2000, Leiden.

Boakye-Boaten, Akwasi. "Cultural Ecology of Asante, 1702–1945." Unpublished PhD diss., University of Ghana, Legon, 1974.

Boaten, K. "A Historical Geography of Kwabere." Unpublished BA thesis, University of Ghana, Legon, [n.d.].

Bredwa-Mensah, Yaw. "Archaeology of Slavery in West Africa." Unpublished conference paper; World Archaeological Congress 4, Cape Town, 10–14 January 1999.

Chouin, Gérard. "Forests of Power and Memory: An Archaeology of Sacred Groves in the Eguafo Polity, Southern Ghana (c. 500–1900 A.D.)." Unpublished PhD diss., Syracuse University, 2009.

Elbl, Ivana. "The Portuguese Trade with West Africa, 1440–1521." Unpublished PhD diss., University of Toronto, 1986.

Kea, Raymond. "Trade, State Formation and Warfare on the Gold Coast, 1600–1826." Unpublished PhD diss., University of London, 1974.

Meecham, Sarah Hand. "Going Over and Coming Back: Gender, Technology, and Household Cidering in England and the Chesapeake 1690 to 1760." Unpublished conference paper; Omohundro Institute of Early American History and Culture Conference, 14–16 June 2002, College Park MD.

Nan, Josée. "Rimadoors: een Afrikaans aandeel in de handel tussen Nederland en de Goudkust, 1602–1872." Unpublished thesis, Universiteit Leiden, 1981.

Piot, Charles . "Production, Exchange and Symbolic Forms among the Kabre." Unpublished PhD diss., Anthropology, University of Virginia, 1986.

Quaye, I. "The Ga and Their Neighbors, 1600–1742." Unpublished PhD diss., University of Ghana, Legon, 1972.

Sanders, James R. "The Political Development of the Fante in the Eighteenth and Nineteenth Centuries: a Study of a West African Merchant Society." Unpublished PhD diss., Northwestern University, 1981.

### III. *Published Primary and Secondary Works*

Abayie Boaten, Nana Akwasi. "Asante: The Perception and the Utilization of the Environment before the Twentieth Century." *Research Review*, new series, 6, 2 (1990), 19–28.

——. "Kumase: Early Settlement up to the End of the 18th Century." *Research Review*, supplement 5 (1990), 1–15.

Abbiw, Daniel K. *Useful Plants of Ghana*. London: Intermediate Technology Publishing, 1990.

Adams, C. D. "Activities of the Danish Botanists in Guinea, 1783–1850." *Transactions of the Historical Society of Ghana* 3, 1 (1957), 30–46.

Adams, Capt. John. *Sketches taken during ten voyages to Africa between the years 1786 and 1800*... London: Frank Cass, 1823.

Affran, Dan Kojo. "Cassava and Its Economic Importance." *Ghana Farmer* 12, 4 (1968), 172–78.

Akyeampong, Emmanuel. *Between the Sea and the Lagoon: An Eco-Social History of the Anlo of Southeastern Ghana c. 1850 to Recent Times*. Athens OH: Ohio University Press / London: James Currey, 2001.

——. *Drink, Power, and Cultural Change: A Social History of Alcohol, c. 1800 to Recent Times*. Portsmouth NH: Heinemann, 1996.

Allsopp, Richard, and Jeannette Allsopp. *Dictionary of Caribbean English Usage*. Oxford: Oxford University Press, 1996.

Allsworth-Jones, P. "The Earliest Human Settlement in West Africa and the Sahara." *West African Journal of Archaeology* 17 (1987), 87–128.

Alpern, Stanley. "Did They or Didn't They Invent It?: Iron in Sub-Saharan Africa." *History in Africa* 32 (2005), 41–94.

——. "The European Introduction of Crops to West Africa." *History in Africa* 19 (1992), 13–43.

——. "The Exotic Plants of Western Africa: Where They Came From and When." *History in Africa* 35 (2008), 63–102.

Amos, Alcione M., and Ebenezer Ayesu, "'I am Brazilian': History of the Tabon, Afro-Brazilians in Accra, Ghana." *Transactions of the Historical Society of Ghana*, new series, 6 (2002), 35–58.

Andah, Bassey. "Agricultural Beginning and Early Farming Communities in West and Central Africa." *West African Journal of Archaeology* 17 (1987), 171–204.

——. "Early Paleolithic in West Africa: The Case of the Asokrochona Coastal Region of Accra Ghana." *West African Journal of Archaeology* 9 (1979), 47–85.

——. "Identifying the Early Farming Traditions in West Africa." In Thurston Shaw, Paul Sinclair, Bassey Andah, and Alex Opoko, eds., *The Archaeology of Africa: Food, Metal, and Towns* (London and New York: Routledge, 1993), 240–54.

Annor-Frempong, C. "A Survey of Cassava Cultivation Practices in Ghana." *Acta horticulturæ* 380 (1994), 216–21.

Anonymous Dutch Manuscript.
see Jones, 1995.

Anquandah, James. "Boyasi Hill: A Kintampo Culture Site." *Sankofa* 2 (1976), 92.

——. *Rediscovering Ghana's Past*. London: Longman, 1982.

Armitage, David, and Michael J. Braddick, eds. *The British Atlantic World, 1500–1800*. Basingstoke and New York: Palgrave Macmillan, 2002.

Austen, Ralph. *African Economic History*. London: James Currey / Portsmouth NH: Heinemann, 1987.

Ayiteh-Smith, E. *Prospects and Scope of Plant Medicine in Heath Care*. Accra: Ghana Universities Press, 1989.
Baesjou, René. An Asante Embassy on the Gold Coast: The Mission of Akyempon Yaw to Elmina, 1869–1872. Leiden: Afrika Studiecentrum, 1979.
Ballong-Wen-Mewuda, J. Bato'ora. "Le commerce portugais des esclaves entre la Côte de l'actuel Nigéria et celle du Ghana moderne aux XVe et XVIe siècles." In Serge Daget, ed., *De traite à l'esclavage: Actes du Colloque International sur la Traite des Noires, Nantes 1985*, 2 vols. Paris: Société française d'histoire d'outre-mer, 1988.
——. *São Jorge da Mina, 1482–1637: La vie d'un comptoir portugais en Afrique occidental*, 2 vols. Lisbon and Paris: Écoles des Hautes Études en Sciences Sociales, collection du Centre d'Études Portugaises: Fondation Calouste Gulbenkian/Commision Nationale pour les Commérations des Décourvertes Portugaises, 1993.
Bannerman, J. Yedu. *Mfantse-Akan mbɛbuɛsm nkyerɛkyɛrmu*. Tema: Hacquason Press, [n.d.].
Barbot, Jean.
see Hair, Jones and Law, 1992.
Bendor-Samuel, John and Rhonda L. Hartell, eds. *The Niger–Congo languages*. Lanham, MD: The University Press of America, 1989.
Bennet, Norman R. and George. E. Brooks, eds. *New England Merchants in Africa: A History Through Documents, 1802 to 1865*. Boston: African Studies Program, Boston University, 1965.
Beoku-Betts, Josephine A. " 'We Got Our Way of Cooking Things': Women, Food, and Preservation of Cultural Identity among the Gullah." *Gender & Society* 9, 5 (1995), 535–55.
Berlin, Ira and Philip D. Morgan, eds. *Civilization and Culture: Labor and the Shaping of Slave Life in the Americas*. Charlottesville: University Press of Virginia, 1993.
Berry, Sara S. *Chiefs Know Their Boundaries: Essays on Property, Power, and the Past in Asante, 1896–1996*. Portsmouth NH: Heinemann, 2001.
Bilby, Kenneth. *True-Born Maroons*. Gainesville: University Press of Florida, 2005.
Blake, J. W., trans. and ed. *Europeans in West Africa, 1450–1560*. London: The Hakluyt Society, 1942.
Bleek, Wolf. "Did the Akan Resort to Abortion in Pre-Colonial Ghana?: Some Conjectures." *Africa* (International African Institute) 60, 1 (1990), 121–31.
Blench, Roger. "Bananas and Plantains in Africa: Reinterpreting the Evidence," *Ethnobotany Research and Applications* 7 (2009), 363–80.
——. "Prehistory of African Ruminant Livestock, Horses and Ponies." In Shaw, *et al.*, *Food, Metals, and Towns*, 71–103.
——. "Unnatural Terms for Natural Things: Comparative Ethnobotany in the Volta Region of Ghana and Its Implications for Historical Linguistics." Unpublished conference paper; Colloquium of African Languages and Linguistics, August 2000, Leiden.
Blench, Roger, Kay Williamson, and Bruce Connell. "The Diffusion of Maize in Nigeria: A Historical and Linguistic Investigation." *Sprache und Geschichte in Afrika* 15, 1 (1994), 9–46.
Boahen, Adu. "Ghana before the Coming of Europeans." *Ghana Social Science Journal* 4, 2 (1977), 93–106.
Boateng, E. A. *A Geography of Ghana*, 2nd ed. Cambridge: Cambridge University Press, 1966.
Bonnat, Marie-Joseph, see Perrot and Van Dantzig, 1994.
Borlaug, Norman E., *et al. Lost Crops of Africa: Grains*. Washington DC: National Academy Press, 1996.
Boserup, Esther. *The Conditions of Agricultural Growth*. Chicago: Aldine, 1965.
Bosman, Willem. *Nauwkeurige beschryving van de Guinese Goud- Tand- en Slave-Kust*. Utrecht, 1704.
Boulègue, Jean. *Les Luso-Africains de Sénégambie*. Lisbon: Ministério da Educação, 1989.

Bowdich, T. E. *Mission from Cape Coast Castle to Ashantee, with a Statistical Account of that Kingdom, and Geographical Notices on Other Parts of the Interior Parts of Africa*. London, 1819.
Bramer, H. "Soils." In J. B. Wills, ed., *Agriculture and Land Use in Ghana* (London: Oxford University Press, 1962), 88–126.
Brásio, António, ed. *Monumenta Missionaria africana: Africa ocidental (1600–1622)*, sér. 1, vol. 1. Lisbon: Agência Geral do Ultramar, Divisão de Publicações e Biblioteca, 1952.
Burkhill, H. M. *The Useful Plants of West Tropical Africa*, 5 vols. Kew: Royal Botanic Gardens, 1985–2000.
Busia, K. A. *The Position of the Chief in the Modern Political System of Ashanti*. London: Oxford University Press, 1951.
Carney, Judith. "African Rice in the Columbian Exchange." *Journal of African History* 42, 3 (2001), 377–96.
——. "The African Origins of Carolina Rice Culture." *Ecumene: A Journal of Cultural Geographies* 7, 2 (2000), 125–49.
——. *Black Rice: The African Origins of Rice Cultivation in the Americas*. Cambridge MA: Harvard University Press, 2001.
——. "The Role of African Rice and Slaves in the History of Rice Cultivation in the Americas." *Human Ecology* 26 (1998), 525–45.
Chenorkian, Robert. "Ivory Coast Prehistory: Recent Developments." *African Archaeological Review* 1 (1983), 127–42.
Chouin, Gérard. "Seen, Said, or Deduced? Travel Accounts, Historical Criticism, and Discourse Theory: Towards an "Archaeology" of Dialogue in Seventeenth-Century Guinea." *History in Africa* 28 (2001), 53–70.
Chouin, Gérard L. and Christopher R. DeCorse. "Prelude to the Atlantic Trade: New Perspectives on Southern Ghana's Pre-Atlantic History (800–1500)." *Journal of African History* 51, 1 (2010), 123–45.
Chrétien, J. P. "Les années de l'éleusine, du sorgho et du haricot dans l'ancien Burundi ecologie et idéologie." *African Economic History* 7 (1979), 75–92.
Christaller, J. G. *A Dictionary of the Asante and Fante Language, Called Tschi*. Basel: Basel Evangelical Missionary Society, 1881, 2nd ed. revised and enlarged, 1933.
——. *Three Thousand Six Hundred Proverbs (from the Asante and Fante Languange)* [originally published as *A Collection of Three Thousand and Six Hundred Tshi Proverbs* (Basel, 1879)], trans. Kofi Ron Lange. Lewiston NY: Edwin Mellon Press, 1990.
Claridge, W. Walton, *A History of the Gold Coast and Ashanti from the Earliest Times to the Commencement of the Twentieth Century*, 2nd ed. London: Cass, 1964.
Clutton-Brock, Juliet. "The Spread of Domestic Animals in Africa." In Shaw, *et al.*, eds., *The Archaeology of Africa: Food, Metal, and Towns*, 61–70.
Coclanis, Peter A. "*Drang nach Osten*: Bernard Bailyn, the World-Island, and the Idea of Atlantic History." *Journal of World History*, 13, 1 (2002), 169–82.
Counihan, Carole M., and Penny Van Esterik, eds. *Food and Culture: A Reader*. New York: Routledge, 1997.
Coursey, D. G. "The Origin and Development of Agriculture in Africa." In Jack Harlan, J. M. J. de Wet, and Ann Stemler, eds., *The Origins of African Plant Domestication*. The Hague: Mouton, 1976.
——. "The Origin and Domestication of Yams in Africa." In Harlan, De Wet, and Stemler, eds., *The Origins of African Plant Domestication*, 383–408.
Crist, Raymond E. "Rice Culture in Spain." *Scientific Monthly* 84, 1 (1957), 66–74.
Crone, G. R., trans. and ed. *The Voyages of Cadamosto and Other Documents*. London: The Hakluyt Society, 1937.
Cruickshank, Brodie. *Eighteen Years on the Gold Coast of Africa, including an Account of the Native Tribes*, 2 vols. London: Hurst and Blackett, 1853.
Curtin, Philip. "Africa in the Wider Monetary World, 1250–1850," In John F. Richards, ed., *Silver and Gold Flows in the Medieval and Early Modern Worlds* (Durham NC: Duke University Press, 1981), 231–68.

——. *The Atlantic Slave Trade: A Census*. Madison: University of Wisconsin Press, 1969.

Daaku, "Kwame Yeboa. "Trade and Trading Patterns of the Akan in the Seventeenth and Eighteenth Centuries." In Claude Meillassoux, ed., *The Development of Indigenous Trade and Markets in West Africa* (London: Oxford University Press, 1971), 168–81.

Dalby, David, and P. E. H. Hair. "A Further Note on the Mina Vocabulary of 1479–80." *Journal of West African Languages* 5, 2 (1968), 129–31.

Dalziel, J. M. *The Useful Plants of West Tropical Africa*. London: Crown Agents, 1937.

Danquah, J. B. "Notes on 'Obroni' and 'Buronya.'" *Transactions of the Gold Coast and Togoland Historical Society* 2 (1952), 71–71.

Danquah, O. A. "Fungi Associated with Discoloured Rice (*Oryza sativa*) Seeds in Ghana." *Ghana Journal of Agricultural Science* 9, 3 (1976), 185–87.

Dapper, Olfert. *Naukeurige beschrijvinge der Afrikaensche gewesten . . .* Amsterdam, 1668.

Da Silva, Joaquim Lino. "O *Zea mays* e os milhos africanos na costa oriental da África: discussão em torno de documentos antigos." *Garcia de Orta, Série de Geografia* 14, nos. 1–2 (1993), 15–40.

Data, Ansu, and R. Porter. "The Asafo System in Historical Perspective." *Journal of African History* 12, 2 (1971), 279–97.

Davies, Oliver. "The Ntereso Culture in Ghana." In B. K. Schwartz and R. A. Dumett, eds., *West African Culture Dynamics: Archaeological and Historical Perspectives*. (The Hague: Mouton, 1980), 205–25.

——. *The Quaternary in the Coastlands of Guinea*. Glasgow: Jackson, 1964.

——. *West Africa before the Europeans*. London: Methuen, 1967.

DeCorse, Christopher. "Culture Contact, Continuity, and Change on the Gold Coast, AD 1400–1900." *African Archaeological Review* 10 (1992), 163–96.

De la Fosse, d'Eustache.
see Escudier, 1992.

De Marée, Johannes Adam. *Reizen op en beschrijving van de Goudkust van Guinea*, 2 vols. 's-Gravenhage and Amsterdam: Gebroeders Van Cleef, 1817–8.

De Marees, Pieter.
see Naber, 1912.
see Van Dantzig and Jones, 1987.

Diamond, Jared. *Guns, Germs, and Steel: the Fates of Human Societies*. New York: Norton, 1997.

Dimmendaal, Gerrit J. *Historical Linguistics and the Comparative Study of African Languages*. Amsterdam and Philadelphia: John Benjamins, 2011.

Doku, E. V. *Cassava in Ghana*. Accra: Ghana Universities Press, 1969.

Doortmont, Michel. "An Overview of Dutch Relations with the Gold Coast in Light of David van Nyendael's Mission to Asante in 1701–02," in Ineke van Kessel, ed., *Merchants, Missionaries & Migrants: 300 years of Dutch-Ghanaian Relations* Amsterdam: KIT Publishers / Accra: Sub-Saharan Publishers 2001), 19–31.

Ekundayo, J. A. "The Production of Pito, A Nigerian Fermented Beverage." *Journal of Food Technology* 4 (1969), 217–25.

Elbl, Ivana. "The Volume of the Early Atlantic Slave Trade, 1450–1521." *Journal of African History* 38, 1 (1997), 31–75.

Eldredge, Elizabeth. "Drought, Famine, and Disease in Nineteenth-Century Lesotho." *Journal of African History* 16 (1987), 61–93.

——. "Sources of Conflict in Southern Africa, c. 1800–30: The 'Mfecane' Reconsidered." *Journal of African History* 33, 1 (1992), 1–35.

Eltis, David, and Philip Morgan and David Richardson. "Agency and Diaspora in Atlantic History: Reassessing the African Contribution to Rice Cultivation in the Americas." *The American Historical Review*, 112, (December 2007), 1329–1358.

Eltis, David, and Stephen D. Behrendt, David Richardson, and Herbert S. Klein. *The Trans-Atlantic Slave Trade: A Database on CD-ROM*. Cambridge and New York: Cambridge University Press, 1999.

Enti, Albert A. *The Rejuvenating Plants of Tropical Africa*. San Francisco: North Scale Institute, 1988.
Escudier, Denis, ed. *Voyage d'Eustache Delafosse sur la côte Guinée au Portugal et en Espange, 1479–81*. Paris: Editions Chandeigne, 1992.
Everts, Natalie. " 'Brought up well according to European Standards:' Helena van der Burgh and Wilhelmina van Naarssen: Two Christian women from Elmina," in Ineke van Kessel, ed., *Merchants, Missionaries & Migrants: 300 years of Dutch-Ghanaian Relations* Amsterdam: KIT Publishers / Accra: Sub-Saharan Publishers 2001), 101–109.
——. "Cherchez la femme: Gender-Related Issues in Eighteenth-Century Elmina." *Itinerario* 20, 1 (1996), 45–57.
Fage, J. D. *Original Sources for Precolonial Western Africa Published in European Languages*, 2nd ed. rev. Madison: African Studies Program, University of Wisconsin-Madison, 1994.
——. "Seventeenth Century Gold Coast." [review of Van Dantzig and Jones, *Pieter de Marees*] *Journal of African History* 29, 2 (1988), 326–28.
——. "Slaves and Society in Western Africa, c. 1445–1700." *Journal of African History* 21, 3 (1980), 289–310.
Fairhead, James, and Melissa Leach. *Misreading the African Landscape: Society and Ecology in a Forest-Savanna Mosaic*. Cambridge: Cambridge University Press, 1996.
Feinberg, Harvey. *Africans and Europeans in West Africa: Elminians and Dutchmen on the Gold Coast during the Eighteenth Century*. Philadelphia: Transactions of the American Philosophical Society, 1989.
Felius, Marleen. *Cattle Breeds—An Encyclopedia*. Doetinchem: Misset, 1995.
Fields, Edda. *Deep Roots: Rice Farmers in West Africa and the African Diaspora*. Bloomington: Indiana University Press, 2008.
Fortes, M. *Kinship and the Social Order*. Chicago: Aldine, 1969.
Foulkes, Arthur. "The Company System in Cape Coast Castle." *Journal of the African Society* 7 (1907–08), 262–77.
Games, Allison. "Atlantic History: Definitions, Challenges, and Opportunities." *American Historical Review* 111, 3 (June 2006), 741–57.
Giblin, James. "Trypanosomiasis Control in African History: An Evaded Issue." *Journal of African History* 30, 1 (1990), 59–80.
Gifford-Gonzalez, Diane, and Olivier Hanotte. "Domesticating Animals in Africa: Implications of Genetic and Archaeological Findings." *Journal of World Prehistory* 24 (2011), 1–23.
Gill, Michael M. "Annotated List of Plantain Cultivars in Ghana." *The Ghana Farmer* 15, 2 (1971), 88–93.
Gledhill, D. *West African Trees*. London: Longman, 1972.
Glick, Thomas F. *Irrigation and Society in Medieval Valencia*. Cambridge: Cambridge University Press, 1970.
Golden, Michael H. N. "Oedematous Malnutrition." *British Medical Bulletin* 54, 2 (1998), 434–44.
Goodman, Major M. "Maize (*Zea mays*)." In N. W. Simmonds, ed., *Evolution of Crop Plants* (New York: Longman, 1976), 128–36.
Goody, Jack. *Cooking, Cuisine and Class: a Study in Comparative Sociology*. Cambridge: Cambridge University Press, 1982.
Gramberg, Jan Simon Gerhardus. *Schetsen van Afrikas Westkust*. Amsterdam: Weijtingh & Brave, 1861.
Green, Sandra E. *Gender, Ethnicity, and Social Change on the Upper Slave Coast*. Portsmouth NH: Heinemann, 1996.
Grigson, Caroline. "An African Origin for African Cattle?" *African Archaeological Review* 9 (1991), 119–44.
Gulbrandsen, O. "On the Problem of Egalitarianism: The Kalahari San in Transition." *Journal of Comparative Social Religion* 10–11 (1991), 12–31.

Haaland, Randi. "Fish, Pots, and Grain: Early and Mid Holocene Adaptations in the Central Sudan." *African Archaeological Review* 10 (1992), 43–64.
Hagan, G. P. "A Note on Akan Colour Symbolism." *Research Review* 7, 1 (1970), 8–14.
Hair, P. E. H. "Collections of Vocabularies of Western Africa before the Polyglotta: a Key." *Journal of African Languages* 5 (1966), 208–17.
——. "A Note on De la Fosse's 'Mina' Vocabulary of 1479–80." *Journal of West African Languages* 3, 1 (1966), 55–7.
——. *The Founding of the Castelo de São Jorge da Mina: An Analysis of the Sources*. Madison: African Studies Program, University of Wisconsin-Madison, 1994.
Hair, P. E. H., Adam Jones and Robin Law, trans. and eds., *Barbot on Guinea: The Writings of Jean Barbot on West Africa*, 2 vols. London: The Hakluyt Society, 1992.
Hall, Robert. "Savoring Africa in the New World." In Herman J. Viola and Carolyn Margolis, eds., *Seeds of Change: A Quincentennial Commemoration* (Washington DC: Smithsonian Institute Press, 1991), 161–77.
Harlan, Jack, and J. M. J. de Wet, and Ann Stemler, eds. *Origins of African Plant Domestication*. The Hague: Mouton, 1976.
——. "Plant Domestication and Indigenous African Agriculture." In Harlan, De Wet, and Stemler, eds., *The Origins of African Plant Domestication*, 3–19.
Harris, M., and E. B. Ross, eds. *Food and Evolution: Toward a Theory of Human Food Habits* (Philadelphia: Temple University Press, 1997), 57–90.
Hauser-Renner, Heinz. "Examining Text Sediments—Commending a Pioneer Historian as an 'African Herodotus': On the Making of the New Edition of C. C. Reindorf's *History of the Gold Coast and Asante*." *History in Africa* 35 (2008), 231–300.
Hawthorne, Walter. *From Africa to Brazil: Culture, Identity, and an Atlantic Slave Trade, 1600–1830*. Cambridge UK: Cambridge University Press, 2010.
——. "Nourishing a Stateless Society during the Slave Trade: The Rise of Balanta Paddy-Rice Production in Guinea-Bissau". *Journal of African History* 42, 1 (2001), 1–24.
——. *Planting Rice and Harvesting Slaves: Transformations along the Guinea-Bissau Coast, 1400–1900*. Portsmouth NH: Heinemann, 2003.
Heintze, Beatrix, and Adam Jones, eds. "European Sources for Sub-Saharan Africa before 1900: Use and Abuse." *Paideuma* 33 (1987) [special issue].
Henige, David. "Gambit Declined: Pervasive Doubt about Systematic Doubt." In Robert Harms, et al., eds., *Paths towards the Past: African Historical Essays in Honor of Jan Vansina* (Atlanta: ASA Press, 1994), 77–95.
——. *Oral Historiography*. London: Longman, 1982.
——. "The Problem of Feedback in Oral Tradition: Four Examples from the Fante Coastlands." *Journal of African History* 14, 2 (1973), 223–35.
Hepper, F. N. *The West African Herbaria of Isert and Thonning: A Taxonomic Revision and Index to the IDC Microfiche*. Kew: Bentham-Maxon Trust and the Royal Botanical Gardens, 1976.
Herbert, Eugenia W. *Red Gold of Africa: Copper in Precolonial History and Culture*. Madison: University of Wisconsin Press, 1984.
Hernæs, Per. *Slaves, Danes, and African Coast Society: The Danish Slave Trade from West Africa and Afro-Danish Relations on the Eighteenth Century Gold Coast*. Trondheim: Department of History, University of Trondheim, 1995.
Heywood, Linda Marinda, and John Kelly Thornton. *Central Africans, Atlantic Creoles and the Foundation of the Americas, 1585–1660*. Cambridge: Cambridge University Press, 2007.
Irvine, F. R. *West African Crops*. Oxford: Oxford University Press, 1974.
Iselin, Regula. "Reading Pictures: On the Value of the Copperplates in the *Beschryvinghe* of Pieter de Marees (1602) as Source Material for Ethnohistorical Research." *History in Africa* 21 (1994), 147–70.
Isert, Paul Erdmann.
see Winsnes, 1992.

Jane, Cecil, ed. *The Journal of Christopher Columbus*. New York: Bramball House, 1960.
Janzen, John, and Steven Feierman, eds. *The Social Basis of Health and Healing in Africa*. Berkeley: University of California Press, 1992.
Jeffreys, M. D. W. "How Ancient is West African Maize?" *Africa* (International African Institute, London) 33 (1963), 115–31.
——. "The Origin of the Portuguese Word Zaburro as Their Name for Maize." *Bulletin de l'Institut Française d'Afrique Noire* 29, ser. B, nos. 1–2 (1957), 111–36.
——. "Pre-Columbian Maize in Africa." *Nature* 172 (1953), 25–35.
——. "Pre-Columbian Negroes in America." *Scientia* 58 (1953), 202–18, and supplement, 113–28.
Jeník, J., and J. B. Hall. "Plant Communities of the Accra Plains, Ghana." *Folia Geobotanica et Phytotaxonomica* 11 (1976), 163–212.
Jenkins, Paul, ed., *Recovery of the West African Past: African Pastors and African History in the Nineteenth Century: C. C. Reindorf and Samuel Johnson*. Basel: Basler Afrika Bibliographien, 1998.
Jennings, D. L. "Cassava (*Manihot esculenta*)." In N. W. Simmonds, ed., *Evolution of Crop Plants* (London: Longman, 1976), 81–84.
Jeppesen, Henrik. "Danske plantageanlæg på Guldkysten, 1788–1850." *Geografisk Tidskrift* 65, 1 (1966), 48–72.
Jones, Adam. "Decompiling Dapper: a Preliminary Search for Evidence." *History in Africa* 17 (1990), 171–209.
——. "Drink Deep, or Taste Not: Thoughts in the Use of Early European Records in the Study of African Material Culture." *History in Africa* 21 (1994), 349–70.
——. "Female Slave-Owners of the Gold Coast: Just a Matter of Money?" In Stephan Palmié, ed., *Slave Cultures and the Cultures of Slavery* (Knoxville: University of Tennessee, 1995), 100–11.
——. "Prostitution, Polyandrie oder Vergewaltigung? Zur Mehrdeutigkeit Europäischer Quellen über der Küste Westafrikas zwischen 1660 und 1860." In Adam Jones, ed., *Außereuropäische Frauengeschichte: Probleme der Forschung* (Pfaffenweiler: Centaurus Verlag, 1990), 123–58.
——. *Zur Quellenproblematik der Geschichte Westafrikas, 1450–1900*. Stuttgart: Franz Steiner Verlag, 1990.
Jones, Adam, trans. and ed. *Brandenburg Sources for West African History, 1680–1700*. Stuttgart: Steiner, 1985.
——. *West Africa in the Mid-Seventeenth Century: An Anonymous Dutch Manuscript*. Atlanta: ASA Press, 1995.
Jones, William O. *Manioc in Africa*. Stanford: Stanford University Press, 1959.
Juhé-Beaulaton, Dominique. "Les jardins des forts Européens de Ouidah: premiers jardins d'essai (XVIIIème siècle)." *Cahiers du Centre de Recherches Africaines* 8 (1994), 83–105.
Justesen, Ole, ed. *Danish Sources for the History of Ghana 1657–1754*, 2 vols. Copenhagen: The Royal Danish Academy of Sciences and Letters, 2005.
Karikari, S. K. "A Note on Plantain (*Musa* AAB Group) and Banana (*Musa* ABB Group) Cultivars in Ghana." *Ghana Journal of Agricultural Science* 4 (1971), 79–85.
——. "Some Taxonomic Assessment of Contribution of *Musa acuminata* and *Musa balbisiana* to the Origins of Plantains and Bananas in Ghana." *Ghana Journal of Agricultural Science* 6 (1973), 9–19.
Kea, Raymond A. "Plantations and Labor in the South-East Gold Coast from the Eighteenth to Mid-Nineteenth Century," In Robin Law, ed., *From Slave Trade to Legitimate Commerce* (Cambridge: Cambridge University Press, 1995), 119–43.
——. "Some Danish Accounts of the 1815–16 Asante Military Expedition on the Gold Coast." *Ghana Studies Newsletter* 10 (1997), 6–9.
——. *Trade, Settlements, Polities in the Seventeenth-Century Gold Coast*. Baltimore: Johns Hopkins University Press, 1982.

Kern, H. ed., *Het Itinerario van Jan Huygen van Linschoten, 1579–1592*, 2 vols. 's-Gravenhage: Martinus Nijhoff, 1910.
Kimble, G. H. T., trans. and ed. *Esmeraldo de situ orbis*. London: The Hakluyt Society, 1937.
Kiple, Kenneth F., and Kriemhild Coneè Ornelas, eds. *The Cambridge World History of Food*, 2 vols. Cambridge: Cambridge University Press, 2000.
Klein, A. Norman. "Slavery and Akan Origins?" *Ethnohistory* 41, 4 (1994), 627–56.
——. "Toward a New Understanding of Akan Origins." *Africa* 66, 2 (1996), 248–73.
Klingelhoffer, Anne. "Agriculture." In David Brokensha, ed., *Akwapim Handbook* (Tema: Ghana Publishing Corporation, 1972), 130–37.
Koeppel, Dan. *Banana: The Fate of the Fruit That Changed the World*. London: Penguin, 2008.
Konotey-Ahulu, F. I. D. "Issues in Kwashiorkor." *The Lancet* 343 (1994), 548.
——. "Kwashiorkor." *British Medical Journal* 302 (1991), 180–81.
Kropp Dakubu, M. E. "A Survey of Borrowed Words in Dangme." *Research Review*, supplement no. 4 (1973), 81–128.
——. *Ga-English Dictionary*. Legon: Institute of African Studies, University of Ghana, 1973.
——. "Notes on the Linguistic Situation on the Coast of Ghana during the Nineteenth Century." *Research Review*, new series, 1, 2 (1985), 192–202.
——. "The Peopling of Southern Ghana: A Linguistic Viewpoint." In Christopher Ehret and Merrick Posnansky, eds., *The Archaeological and Linguistic Reconstruction of African History* (Berkeley: University of California Press, 1982), 245–55.
Kupperman, Karen Ordahl. "Fear of Hot Climates in the Anglo-American Colonial Experience." William and Mary Quarterly, Third Series, 41, 2 (April 1984), 213–240.
La Fleur, J. D., trans. and ed. *Pieter van den Broecke's Journal of Voyages to Cape Verde, Guinea and Angola, 1605–1612*. London: The Hakluyt Society, 2000.
——. "Atlantic-era Experimentation with Plantains on the Gold Coast, West Africa" (paper presented at the annual meeting of African Studies Association conference, Washington, DC; 6 December 2002).
——. "The Culture of Crops on the Gold Coast (West Africa) from the Earliest Times to circa 1850." PhD diss. in History, University of Virginia, 2003.
——. "Plants: Imported Species" in John Middleton and Joseph Miller, eds., *New Encyclopedia of Africa*, 2nd ed. New York: Scribner, 2008. Vol. 4, 158–59.
Lange, Kofi Ron.
See Christaller, 1933.
Langer, William L. "Europe's Initial Population Explosion." *American Historical Review* 69, 1 (1963), 1–17.
Larson, Pier M. *History and Memory in the Age of Enslavement: Becoming Merina in Highland Madagascar, 1770–1822*. Portsmouth NH: Heinemann, 2000.
Laudan, Rachel. "Birth of the Modern Diet." *Scientific American* 283, 2 (2000), 76–81.
Law, Robin. *The Kingdom of Allada*. Leiden: Research School CNWS, 1997.
Law, Robin, ed., *The English in West Africa: The Local Correspondence of the Royal African Company of England, 1681–1699*, multiple vols. Oxford: Oxford University Press, for the British Academy, 1997–continuing.
Law, Robin, and Paul Lovejoy. "The Changing Dimension of African History: Reappropriating the Diaspora." In Simon McGrath, *et al., Rethinking the African Diaspora* (Edinburgh: University of Edinburgh CAS, 1997), 181–200.
Layton, R., R. Foley, and E. Williams. "The Transition between Hunting and Gathering and the Specialized Husbandry of Resources: a Socio-Ecological Approach." *Current Anthropology* 32 (1991), 255–75.
Lazzarini-Viti, Veronica. "Western Foods and Traditional Diet in Ghana." *Paideuma* 24 (1978), 103–09.
Leers, Arnout. *Pertinente beschryvinghe van Africa*… Rotterdam, 1665.

Lindsay, Lisa A. "'To Return to the Bosom of their Fatherland': Brazilian Immigrants in Nineteenth-Century Lagos." *Slavery and Abolition* 15, 1 (1994), 22–50.

MacDonald, Kevin C. "Korounkorokalé Revisited: the 'Pays mande' and the West African Microlithic Technocomplex." *African Archaeological Review* 14, 3 (1997), 195–96.

Maier, D. J. E. "Military Acquisition of Slaves in Asante." In David Henige and T. C. McCaskie, eds., *West African Economic and Social History: Studies in Memory of Marion Johnson* (Madison: African Studies Program, University of Wisconsin, 1990), 119–32.

——. "Nineteenth-Century Asante Medical Practices." *Comparative Studies in Society and History* 21, 1 (1979), 63–81.

Makepeace, Margaret, ed. *Trade on the Guinea Coast, 1657–1666: The Correspondence of the English East India Company*. Madison: African Studies Program, University of Wisconsin-Madison, 1991.

Maley, J. "Le cadre paléoenvironmental des refuges forestiers africains: Quelque données et hypothèses." In L. J. K. van den Maesen, *et al.*, eds., *The Biodiversity of African Plants* (Dordrecht: Klumer Academic, 1996), 519–35.

Mandala, Elias C. *Work and Control in a Peasant Economy: A History of the Lower Tchiri Valley in Malawi, 1859–1960*. Madison: University of Wisconsin Press, 1990.

Manning, Patrick. *Slavery and African Life*. Cambridge: Cambridge University Press, 1990.

Marzouk, Yasmine, ed., *Génie rural et génie paysan: sociétés rurales et techniques hydraliques en Afrique*. Paris: Éditions de l'École des Hautes Études en Sciences Sociales, 1989.

Mauny, Raymond. *Tableau géographique de l'Ouest africain au moyen âge d'après les sources écrites, la tradition et l'archéologie*. Dakar: l'Institut français d'Afrique noire, 1961.

McCann, James C. "Agriculture and African History." *Journal of African History* 32, 2 (1991), 507–13.

——. "Climate and Causation in African History." *International Journal of African Historical Studies* 32, 2–3 (1999), 261–79.

——. *Green Land, Brown Land, Black Land: An Environmental History of Africa, 1800–199*. Portsmouth NH: Heinemann, 1999.

——. *Maize and Grace: Africa's Encounter with a New World Crop*. Cambridge MA: Harvard University Press, 2005.

McCaskie, T. C. "Accumulation, Wealth, and Belief in Asante History." *Africa* 53, 1 (1983), 23–43.

——. "Innovational Eclecticism: the Asante Empire and Europe in the Nineteenth Century." *Comparative Studies in Society and History* 14, 1 (1972), 30–45.

——. "Nananom Mpow of Mankessim: An Essay in Fante History." In Henige and McCaskie, eds., *West African Economic and Social History*, 133–50.

——. *State and Society in Pre-Colonial Asante*. Cambridge: Cambridge University Press, 1995.

McGee, Harold. *On Food and Cooking: the Science and Lore of the Kitchen*. New York: Charles Scribner's Sons, 1984.

McIntosh, Roderick J. "The Pulse Model: Genesis and Accommodation of Specialization in the Middle Niger." *Journal of African History* 34, 2 (1993), 194–205.

McIntosh, Roderick, and Susan McIntosh. "Recent Archaeological Research and Dates from West Africa." *Journal of African History* 28, 1 (1987), 1–25.

McIntosh S. K., and R. J. McIntosh. "West African Prehistory (from 10,000 BC to AD 1,000)." *American Scientist* 69 (1981), 602–13.

McIntosh, Wm. Alex, and Mary Zey. "Women as Gatekeepers of Food Consumption: A Sociological Critique." *Food and Foodways* 3, 4 (1989), 317–21.

Mensah, P., A. M. Tomkins, B. S. Drasar, T. J. Harrison. "Antimicrobial Effect of Fermented Ghanaian Maize Dough." *Journal of Applied Bacteriology* 70 (1991), 203–10.

Meredith, Henry. *An Account of the Gold Coast of Africa, with a Brief History of the African Company*. London: Longman, 1812.

Metcalf, George. "Microcosm of Why Africans Sold Slaves: Akan Consumption Patterns in the 1770s." *Journal of African History* 28, 3 (1987), 377–94.

Miller, Duncan E., and Nikolaas J. van der Merwe. "Early Metal Working in Sub-Saharan Africa: A Review of Recent Research." *Journal of African History* 35, 1 (1994), 1–35.

Miller, Joseph C., ed. *The African Past Speaks: Essays on Oral Tradition and History.* Folkestone: Wm. Dawson and Sons, 1980.

——. "History in Africa / Africa and History." *American Historical Review* 104, 1 (February 1999), 1–32.

——. "Retention, Re-Invention, and Remembering: Restoring Identities Through Enslavement in Africa and Under Slavery in Brazil," in José C. Curto and Paul E. Lovejoy, eds., *Enslaving Connections: Changing Cultures of Africa and Brazil during the Era of Slavery* (Amherst NY: Prometheus/Humanity Books, 2003), pp. 81–121.

——. "The Significance of Drought, Disease, and Famine in the Agriculturally Marginal Zones of West-Central Africa." *Journal of African History* 23, 1 (1982), 17–61.

——. *The Problem of Slavery as History: A Global Approach.* New Haven: Yale University Press, 2012.

——. *Way of Death: Merchant Capitalism and the Angolan Slave Trade, 1730–1830.* Madison: University of Wisconsin Press, 1988.

Mindzie, Christophe Mbida, *et al.* "First Archaeological Evidence of Banana Cultivation in Central Africa During the Third Millennium Before Present." *Vegetation History and Archaeobotany* 10 (2001), 1–6.

——. "The initial history of bananas in Africa. A reply to Vansina." *Azania* 40 (2005), 128–35.

——. "Yes, There Were Bananas in Cameroon More than 2000 Years Ago." *infoMusa: The International Journal of Banana and Plantain* 13, 1 (2004), 40–42.

Mintz, Sidney. *Sweetness and Power: The Place of Sugar in Modern History.* New York: Penguin Books, 1985.

Miracle, Marvin P. "Interpretation of Evidence on the Introduction of Maize into West Africa." *Africa* (International African Institute, London) 33, 1 (1963), 132–35.

——. "The Introduction and Spread of Maize in Africa." *Journal of African History* 6, 1 (1965), 39–55.

——. *Maize in Tropical Africa.* Madison: University of Wisconsin Press, 1966.

Monod, T., and A. Teixeira da Mota, and R. Mauny, trans. and eds. *Description de la côte occidentale d'Afrique (Sénégal au Cap de Monte, archipels) par Valentim Fernandes.* Bissau: Centro de Estudos da Guiné Portuguesa, 1951.

Monrad, Rev. Hans Christian. *Bidrag til en skildring af Guinea-kysten og dens Indbyggere.* Copenhagen: A. Seidelin, 1822.

Morgan, Philip D. *Slave Counterpoint: Black Culture in the Eighteenth-Century Chesapeake and Lowcountry.* Chapel Hill: University of North Carolina Press, for the Omohundro Institute of Early American History and Culture, Williamsburg VA, 1998.

Muzzolini, A. "Les climats au Sahara et sur ses bordures, du Pléistocène final l'aride actuel." *Empúries* 47 (1985), 8–27.

Naber, S. P. l'Honoré, ed. *Beschryvinge ende historische verhael vant Gout Koninckrijck van Gunea.* 's-Gravenhage: Martinus Nijhoff, 1912.

Neumanm, Katharina, and Elizabeth Hildebrand. "Early Bananas in Africa: The State of the Art," *Ethnobotany Research and Applications* 7 (2009), 353–62.

Newton, L. E., and S. R. J. Woodell. "A Newly Discovered Site for the Kintampo 'Neolithic' Cultural Tradition." *Sankofa* 2 (1976), 19–22.

Nørregård, Georg. *Danish Settlements in West Africa, 1658–1850.* Boston: Boston University Press, 1966.

Northrup, David. *Africa's Discovery of Europe, 1450–1850.* Oxford: Oxford University Press, 2002.

Nuijten E, and R. van Treuren, P. C. Struik, A. Mokuwa and F. Okry, et al., "Evidence for the Emergence of New Rice Types of Interspecific Hybrid Origin in West African Farmers' Fields." *PLoS ONE* 4, 10 (2009).

Nygaard, S. E., and M. R. Talbolt. "Stone Age Archaeology and Environment on the Southern Accra Plains, Ghana." *Norwegian Archaeological Review* 17 (1984), 19–38.
Ofori-Sarpong, E. "Rainfall Reliability over Ghana." *Journal of Eastern African Research and Development* 21 (1991), 93–106.
Ohadike, D. C. "The Influenza Pandemic of 1918–1919 and the Spread of Cassava Cultivation on the Lower Niger: A Study in Historical Linkages." *Journal of African History* 22, 3 (1981), 379–92.
Osseo-Asare, Fran. "We Eat First with Our Eyes: on Ghanaian Cuisine." *Gastronomica*, 2, 1 (2002), 49–57.
Øvrelid, Jotein, ed. and trans. *Ferd til of fra Guinea*. Oslo: Fonna Forlag 1969.
Palmié, Stephan. "Ackee and Saltfish vs. Amalá con Quimbombó?: a Note on Sidney Mintz' Contribution to the Historical Anthropology of African America Cultures." *Journal de le société des américanistes* 91–92 (2005), 89–122.
——. "Creolization and Its Discontents." *Annual Review of Anthropology* 35 (2006): 433–56.
——. "Intangible Cultural Property, Semiotic Ideology, and the Vagaries of Ethnoculinary Recognition." *African Arts* 42, 4 (Winter 2009), 54–61.
Papadakis, J. *Crop Ecologic Survey in West Africa.* Rome: FAO/UN, 1966.
Parker, John. "The Cultural Politics of Death and Burial in Early Colonial Accra." In David M. Anderson and Richard Rathbone, eds., *Africa's Urban Past* (Portsmouth NH: Heinemann / Oxford: James Currey, 2000), 205–21.
——. *Making the Town: Ga State and Society in Early Colonial Accra.* Portsmouth NH: Heinemann, 2000.
Perrot, Claude-Hélène. "Semailles et moissons dans la région d'Assinie vers 1700." *Journal des Africanistes* 59 (1990), 9–25.
Perrot, Claude-Hélène, and Albert van Dantzig, eds. *Marie-Joseph Bonnat et les Ashanti, journal 1869–1874*. Paris: Musée de l'Homme/Société des Africanistes, 1994.
Pilcher, Jeffrey M. *¡Que vivan los tamales!: Food and the Making of Mexican Identity*. Albuquerque: University of New Mexico Press, 1998.
Piot, Charles. "Of Slaves and the Gift: Kabre Sale of Kin During the Era of the Slave Trade." *Journal of African History* 37, 1 (1996), 31–49.
Piperno, Delores R. *Phytolith Analysis: An Archaeological and Geological Perspective.* San Diego: Academic Press, 1988.
——. *Phytoliths: A Comprehensive Guide for Archaeologists and Paleoecologists.* Lantham MD: AltaMira Press, 2006.
Portères, Roland, "Les appelations des céréales en Afrique IX." *Journal d'agriculture tropicale et de botanique appliquée* 6, nos. 6–7 (1959), 189–233.
——. "L'introduction du maïs en Afrique." *Journal d'agriculture tropicale et botanique appliquée* 2, nos. 5–6 (1955), 221–31.
——. "Vieilles agricultures de l'Afrique intertropicale," *L'Agronomie Tropicale* 5, (1950), 489–507.
Posnansky, Merrick. "Archaeological and Linguistic Reconstruction in Ghana." In Christopher Ehret and Merrick Posnansky, eds., *The Archaeological and Linguistic Reconstruction of African History* (Berkeley: University of California Press, 1982), 256–65.
——. "Archaeology, Technology, and Akan Civilization." *Journal of African Studies* 2, 1 (1975), 24–38.
——. "Early Agricultural Societies in Ghana." In J. Desmond Clark and Steven A. Brandt, eds., *From Hunters to Farmers: the Causes and Consequences of Food Production in Africa* (Berkeley: University of California Press, 1984), 147–51.
——. "Review Article: Yams and the Origins of West African Agriculture [review of D. G. Coursey, *Yams* (London: Longmans, 1967)]." *Odu*, new series, 1 (1969), 101–07.
Price, Richard. *Alabi's World.* Baltimore and London: The Johns Hopkins University Press, 1990.
——. *First Time: The Historical Vision of an Afro-American People*. Baltimore and London: The Johns Hopkins University Press, 1983.

——. "The Miracle of Creolization: A Retrospective." *New West Indian Guide* 75 (2001), 35–64.

——. "Subsistence on the Plantation Periphery: Crops, Cooking, and Labour among Eighteenth-Century Suriname Maroons". In Ira Berlin and Philip D. Morgan, eds., *The Slave's Economy: Independent Production by Slaves in the Americas* [special issue of *Slavery and Abolition* 12 (1991)] (London, Frank Cass, 1991), 107–27.

——. *Travels with Tooy: History, Memory, and the African American Imagination* (Chicago: University of Chicago Press, 2008).

Purseglove, J. W. "The Origin and Migrations of Crops in Tropical Africa." In Harlan, De Wet, and Stemler, eds., *Origins of African Plant Domestication*, 291–309.

Quartey, S. M. "Development of the Human Geography of Ahantaland before 1600." *Ghana Social Science Journal* 4, 1 (1977), 117–27.

Rask, Rev. Johannes. *En kort og sandferdig Rejse-beskrivelse til of fra Guinea*. Trondheim, 1754.

Ratelband, K., ed. *Vijf Dagregisters van het Kasteel São Jorge da Mina (Elmina) aan de Goudkust (1645–1647)*. 's-Gravenhage: Martinus Nijhoff, 1953.

Rathbone, Richard. "The Gold Coast, the Closing of the Atlantic Slave Trade, and Africans of the Diaspora." In Palmie, ed., *Slave Cultures and the Cultures of Slavery*, 55–66.

Rattray, R. S. *Ashanti*. Oxford: Oxford University Press, 1923.

——. *Ashanti Law and Constitution*. Oxford: Clarendon Press, 1929.

Reese, Ty M. "Eating Luxury: Fante Middlemen, British Goods, and Changing Dependencies on the Gold Coast." *The William and Mary Quarterly* 66, 4 (2009), 851–72.

Reid, Richard. "Past and Presentism: The 'Precolonial' and the Foreshortening of African History." *Journal of African History* 52, 1 (2011), 135–55.

Reindorf, C. C. *History of the Gold Coast and Asante, Based on Traditions, and Historical Facts*... Basel: Basel Evangelical Missionary Society, 1895.

Richards, Audrey. *Hunger and Work in a Savage Tribe*. London: G. Routledge & Sons, 1932.

Rindos, D. *The Origins of Agriculture*. Orlando: Academic Press, 1984.

Robertshaw, Peter. "Sibling Rivalry? The Intersection of Archaeology and History," *History in Africa* 27 (2000), 261–86.

Robertson, George A. *Notes on Africa, particularly those parts which are situated between Cape Verde and the River Congo*... London: Sherwood, Neely and Jones, 1819.

Rodney, Walter. *How Europe Underdeveloped Africa*. London: Bogle-l'Ouverture Publications, 1972.

Roe, Daphne A. *A Plague of Corn: A Social History of Pellagra*. Ithaca: Cornell University Press, 1973.

Rømer, Ludewig.
see Winsnes, 2000.

Rossel, Gerda. "The Diffusion of Plantain (*Musa* spp. AAB) and Banana (*Musa* sp. AAA) in Africa: A Case for Linguists, Taxonomists, and Historians, Focused on Nigerian Crop Names." In Richard E. Leakey and L. Jan Slikkerveer, eds., *Origins and Development of Agriculture in East Africa: The Ethnosystems Approach to the Study of Early Food Production in Kenya* (Ames: Iowa State University, 1991), 120–60.

——. *Taxonomic-Linguistic Study of Plantain in Africa*. Leiden: Research School CNWS Publications, 1998.

Rowland, J. R. J. *Dryland Farming in Africa*. London: Macmillan, 1993.

Scapp, Ron, and Brian Seitz, eds. *Eating Culture*. New York: State University of New York Press, 1998.

Schaible, Ulrich E. and Stefan H. E. Kaufmann. "Malnutrition and Infection: Complex Mechanisms and Global Impacts." *Public Library of Science* 4, 5 (May 2007), 806–12.

Schaumloeffel, Marco Aurelio. *Tabom: The Afro-Brazilian Community in Ghana*. Self-published, 2009.

Schmidt, Peter R., ed. *The Culture and Technology of African Iron Production*. Gainesville: University Press of Florida, 1996.

Schoenbrun, David. *A Green Place, A Good Place: Agrarian Change, Gender, and Social Identity in the Great Lakes Region to the 15th Century*. Portsmouth NH: Heinemann, 1998.
——. "We Are What We Eat: Ancient Agriculture Between the Great Lakes." *Journal of African History* 34, 1 (1993), 1–31.
Sefa-Dedeh, S. "Process and Product Characterics of *agbelima*, a Fermented Cassava Product." *Tropical Science* 35 (1995), 359–64.
Sen, Amartya. *Poverty and Famines: An Essay on Entitlement and Deprivation*. Oxford: Oxford University Press, 1981.
Shinnie, P. L., and F. J. Keese. *Archaeology of Gonja, Ghana: Excavations at Daboya*. Calgary: University of Calgary Press, 1989.
Simmonds, N. W., ed. *Evolution of Crop Plants*. London: Longman, 1976.
Siraisi, Nancy G. *Medieval and Early Renaissance Medicine: An Introduction to Knowledge and Practice*. Chicago: University of Chicago Press, 1990.
Sowunmi, M. A. "The Environment: Present and Past of West Africa." *West African Journal of Archaeology* 17 (1987), 41–51.
Stahl, Ann Brower. "Innovations, Diffusion, and Culture Contact: The Holocene Archaeology of Ghana." *Journal of World Prehistory* 8, 1 (1994), 51–112.
——. "Reinvestigation of Kintampo 6 Rock Shelter, Ghana: Implications for the Nature of Culture Change." *African Archaeological Review* 3 (1985), 143–44.
——. "The Archaeology of Global Encounters Viewed from Banda, Ghana." *African Archaeological Review* 16, 1 (1999), 5–12.
Stanton, Jennifer. "Listening to the Ga: Cicely Williams' Discovery of Kwashiorkor on the Gold Coast." In Lawrence Conrad, and Anne Hardy, eds., *Women and Modern Medicine* (Amsterdam: Rodopi, 2001), 149–171 [special issue of *Clio Medica: Perspectives in Medical Humanities* 6 (2001)].
Super, J. C. *Food, Conquest, and Colonization in Sixteenth-Century Mexico*. Albuquerque: University of New Mexico Press, 1988.
Taylor, C. J. *Synecology and Silviculture in Ghana*. London: Thomas Nelson, 1959.
Teixeira da Mota, A., and António Carreira, "*Mihlo Zaburro* and *Mihlo Maçaroca* in Guinea and in the Islands of Cabo Verde." *Africa* (International African Institute, London) 36, 1 (1966), 73–84.
Teixeira da Mota, A., and P. E. H. Hair and L. Bourdon, trans. and eds. *A. Donehla, Descrição da Serra Leoa e dos rios de Guiné do Cabo Verde, 1625*. Lisbon: Centro de Estudos de Cartografia Antiga, Secção de Lisboa, 1977.
Tezenas du Montcel, Hugues. *La Bananier plantain*. Paris: Maisonneuve and Larose, 1987.
Thornton, John K. *Africa and Africans in the Making of the Atlantic World, 1400–1800*, 2nd ed. Cambridge UK / New York: Cambridge University Press, 1998.
——. *Warfare in Atlantic Africa, 1500–1800*. London: UCL Press, 1999.
Tilleman, Eric.
see Winsnes, 1994.
Van Dantzig, Albert. "English Bosman and Dutch Bosman: A Comparison of Texts." *History in Africa* 2 (1975) to 11 (1984).
——. *Forts and Castles of Ghana*. Accra: Sedco Publishing Ltd., 1980.
Van Dantzig, Albert, trans. and ed. *The Dutch and the Guinea Coast, 1674–1742*. Accra: Ghana Academy of Arts and Sciences, 1979.
Van Dantzig, Albert, and Adam Jones, trans. and eds. *Pieter de Marees, Description and Historical Account of the Gold Kingdom of Guinea, 1602*. Oxford: Oxford University Press and the British Academy, 1987.
Vansina, Jan. "Bananas in Cameroon c. 500 BCE? Not Proven." *Azania* 38 (2003), 174–76.
——. "The Doom of Early African History?" *History in Africa* 24 (1997), 337–43.
——. "Finding Food and the History of Precolonial Equatorial Africa: a Plea." *African Economic History* 7 (1979), 9–20.
——. "Historians, Are Archaeologists Your Siblings?" *History in Africa* 22 (1995), 369–408.
——. *Living with Africa*. Madison: University of Wisconsin Press, 1994.

——. *Oral Tradition as History*. Madison: University of Wisconsin Press, 1985.
——. *Paths in the Rainforests*. Madison: University of Wisconsin Press, 1988.
——. "The Power of Systematic Doubt in Historical Enquiry." *History in Africa* 1 (1974), 109–27.
Van den Broecke, Pieter.
see La Fleur, 2000.
Van IJzendoorn, P. *De Goudkust: beknopt overzigt van de aangelegenheden betrekkelijk de kust van Guinea, en aldaar van de Goudkust in het bijzonder*. 's-Gravenhage and Amsterdam: Gebroeders Van Cleef, 1817.
Vives, Jamie Vicens. *Economic History of Spain*, trans. by Frances M. López-Morillas. Princeton: Princeton University Press, 1969.
Vogt, John. "The Early São Tomé-Príncipe Slave Trade with Mina, 1500–1540." *International Journal of African Historical Studies* 6, 3 (1973), 453–67.
——. "Private Trade and Slave Sales at São Jorge da Mina: a Fifteenth-Century Document." *Transactions of the Historical Society of Ghana* 15, 1 (1974), 103–10.
——. *Portuguese Rule on the Gold Coast, 1449–1682*. Athens: University of Georgia Press, 1979.
Walker, H. P. "Weather and Climate." In J. B. Willis, ed., *Agriculture and Land Use in Ghana* (London: Oxford University Press, 1962), 7–50.
Watson, Andrew M. *Agricultural Innovation in the Early Islamic World*. Cambridge: Cambridge University Press, 1983.
Wilks, Ivor. *Forests of Gold*. Athens: University of Georgia Press, 1993.
——. "The Forest and the Twis." *Journal des Africanistes* 75, 1 (2005), 19–75.
——. "Land, Labour, Capital and the Forest Kingdom of Asante: A Model of Early Change." In J. Friedman and M. J. Rowlands, eds., *The Evolution of Social Systems* (Pittsburgh: University of Pittsburgh Press, 1978), 487–534.
——. " 'Slavery and Akan Origins?': A Reply." *Ethnohistory* 41, 4 (Fall 1994), 661.
——. "The State of the Akan and the Akan States: A Discursion." *Cahiers d'études africaines* 22, 3–4 (nos. 87–88) (1982), 231–49.
Willet, Frank. "The Introduction of Maize into West Africa: An Assessment of Recent Evidence." *Africa* (International African Institute, London) 32, 1 (1962), 1–13.
Williams, C. D. "A Nutritional Disease of Childhood Associated with a Maize Diet." *Archives of Disease in Childhood* 8, 3 (1933), 423–33.
——. "Kwashiorkor, A Nutritional Disease in Children Associated with a Maize Diet." *The Lancet* 266 (1935), 1151–2.
Winsnes, Selena Axelrod, trans. and ed., *A Reliable Account of the Coast of Guinea (1760) by Ludewig Ferdinand Rømer*. Oxford: Oxford University Press for the British Academy, 2000.
——. *Eric Tilleman: A Short and Simple Account of the Country of Guinea and Its Nature*. Madison: African Studies Program, University of Wisconsin Press, 1994.
——. *Paul Erdmann Isert's Letters on West Africa and the Slave Trade*. Oxford: Oxford University Press, for the British Academy, 1992.
Woodell, S. R. J., and L. E. Newton. "A Standard-Winged Night-Jar Breeding in the Forest Zone of Ghana." *Nigeria Field* 40 (1975), 169–71.
Wrigley, E. A. *Population and History*. New York: Weidenfeld and Nicholson, 1969.
Yarak, Larry W. "Slavery and the State in Asante History." In John Hunwick and Nancy Lawler, eds., *The Cloth of Many Colored Silks: Papers on History and Society Ghanaian and Islamic in Honor of Ivor Wilks* (Evanston: Northwestern University Press, 1997), 223–40.
——. "West African Coastal Slavery in the Nineteenth Century: The Case of the Afro-European Slaveowners of Elmina." *Ethnohistory* 36 (1989), 44–60.
Yentsch, Anne. "Hot, Nourishing, and Culturally Potent: The Transfer of West African Cooking Traditions to the Chesapeake." *Sage: A Scholarly Journal on Black Women* 9, 2 (1995), 15–29.

Zimmermann, Rev. Johannes. *A Grammatical Sketch of the Akra or Ga language . . . a vocabulary of the same, with an appendix on the Adanme dialect*, 2 vols. Stuttgart: Basel Missionary Society, 1858.

Zutshi, Chitralekha. "Debating the Past: Academic and Popular Histories in India." *AHA Perspectives* 47, 9 (December 2009).

# INDEX

*Plants, Animals, and Culinary Items*

## *General Subjects*

*Select Geographical Items*

www.ingramcontent.com/pod-product-compliance
Lightning Source LLC
LaVergne TN
LVHW092117300926
849161LV00002B/160